W9-AHI-048

FRAGILE DOMINION

FRAGILE DOMINION

Complexity and the Commons

SIMON A. LEVIN

HELIX BOOKS

PERSEUS BOOKS
Reading, Massachusetts

Helix Books

Fragile Dominion is part of a continuing series of books based on the Stanislaw M. Ulam Lectures given annually at the Santa Fe Institute in Santa Fe, New Mexico.

Library of Congress Catalog Card Number: 99-61538

Perseus Books is a member of the Perseus Books Group

Jacket design Ella Hanna
Text design by Heather Hutchison
Set in 11-point Galliard

1 2 3 4 5 6 7 8 9 10—03 02 01 00 99
First printing

Find Helix Books on the World Wide Web at http://www.aw.com/gb/

To the memory of my parents, and to Carole—
who has been by my side through everything

CONTENTS

ACKNOWLEDGMENTS

Writing this book has been a joy from beginning to end, the chance to let flow ideas that have been developing and rattling around in my brain for a very long time. Writing the acknowledgments is also a pleasure, but it is in many ways more daunting. How do I thank all of those people who have supported me, fed and critiqued my ideas, and contributed so much to the development of the themes of this book? As with any complex adaptive system, my understanding of and perspectives on the world have been shaped by accidents of history that guided me down certain intellectual pathways. Given the pervasive influence of chance encounters, I feel incredibly fortunate to have interacted with such a remarkable and brilliant stream of people. As my wife, Carole, taught me in our courtship, we are the sum of our contacts with others; in this book, I indeed feel that I am simply channeling the insights I have derived from others. I hope that they all will recognize their input into this book, and will feel that it has been put to good purpose.

My thanks begin with my parents, who encouraged me at every stage of my development. My father transferred to me early on a deep love for science and a need to make one's work serve the purposes of humanity. My mother, who lived until I had finished Chapter 6, asked me every time we spoke how the book was coming. Already ninety-one, she had little idea what the book was about, but she was sure that it was a good thing to do. She knew that that was so because she heard it from my brother, who always seemed to take more pleasure in my accomplishments than in his own considerable ones. He was my constant adviser as I sought my way in science, and his intellectual values reassert themselves in all my explorations.

Carole has been my love and support for most of my life. She has remained apparently interested in everything that I do, simultaneously encouraging and stimulating without ever giving me blanket approval. She has kept me intellectually honest in my work, while giving me enough positive feedback to spur me on. I look for her in my every audience, as she looks for me in hers. Her greatest gifts to me were our children, Jacob and Rachel, who both have been the joys of our lives and have enriched us with their choice of spouses—Karen and Eric. Jacob and Karen's two miracles, Sarah and Ezekiel, continue our good fortune. I have enjoyed my children not only as distractions but also for the scientific exchanges and criticisms they have offered and the inspiration their own careers have given me.

Amy Bordvik has been fantastic in producing this book. I could not have done it without her. She gave me feedback and uncomplainingly dealt with each day's changes of direction and new tasks. Her own sense of style and beauty has helped me often, but mostly I am grateful for her spectacular dedication and hard work. I know that she is as glad as I am to see this work done.

Numerous people have provided invaluable feedback, and critical reading, of various versions of this book. A few deserve special thanks for the very careful and considered comments on numerous earlier versions. In addition to Carole and Amy, I appreciate especially the efforts of four people—Colleen Martin, Ann Kinzig, John Miller, and Amanda Bichsel Cook, my editor at Perseus. Each has mixed positive and negative feedback in ways that have kept me going, while improving the text immeasurably. Along the way, many others have looked critically at particular chapters, and I am especially grateful to Jacob Levin, Jack Repcheck, Jonathan Dushoff, Peter Grant, Karin Limburg, and Dan Rubenstein for specific suggestions. Linda Buttel, Doug Deutschman, Leila Hadj-Chikh, and Lee Worden all helped with the production of figures, under the leadership of Amy Bordvik.

My students and postdocs and my colleagues and collaborators have all contributed more to me than I have to them. I cannot distinguish among the former any more than a father can among his children, and only with trepidation do I endeavor to identify colleagues who have had a special influence, in the certainty that in doing so I will fail to mention others that played important roles. Among my collaborators, however, I would be remiss if I did not mention the early influence of my mentors, Monroe H. Martin and George B. Dantzig; of Bob Whittaker, Dick Root, Bob Paine, Bob May, Lee

Segel, Danny Cohen, David Pimentel, and David Block in the 1970s and 1980s; and of Stu Kauffman, Akira Okubo, Linda Buttel, Rick Durrett, Don Ludwig, Colin Clark, John Steele, Zack Powell, Hal Mooney, Steve Pacala, Jane Lubchenco, Paul Ehrlich, and Glenn Flierl in more recent years. Many of my postdocs and graduate students belong equally on that list, but I do not have the right to leave any of them out, so apologize to those who know that they deserve mention. Some of their names may be found in the pages of this book, and all could have been. The greatest joy of being an academic is the constant renewal, challenge, and learning that comes from interacting with one's students, at all levels, and I have been blessed with wonderful ones.

Several institutions have provided not only the place to develop my ideas in freedom but extraordinary stimulation. Cornell University, where I was on the faculty for twenty-seven years, and Princeton University deserve their reputations as among the world's greatest universities, and I am deeply indebted to the wonderful colleagues they provided me. More recently, I have been blessed with the opportunity to grow in the environment of the Santa Fe Institute, an incredibly stimulating assemblage of brilliant individuals of insatiable curiosity, and the Beijer International Institute of Ecological Economics in Stockholm. Especially influential there have been the director, Karl-Göran Mäler; my predecessor as chairman of the board, Partha Dasgupta; and stimulating colleagues such as Kenneth Arrow, Robert Solow, Paul Ehrlich, and Buzz Holling. I have been fortunate as well to spend sabbatical or other periods at some of the world's greatest centers of learning—Oxford University, the Weizmann Institute, Stanford University, the University of California at Berkeley, the University of Washington, the University of British Columbia, and the University of Kyoto; I acknowledge my debt to all of them, as well as to my alma maters—the Johns Hopkins University and the University of Maryland.

Finally, I am indebted to various foundations and agencies for their support of my work over the years, but especially to the Mellon Foundation and its incomparable talent scout, Bill Robertson; to the Sloan Foundation, and the innovative ideas of Jesse Ausubel; and to the Office of Naval Research, through its University Research Initiative Program award to the Woods Hole Oceanographic Institution.

Invaluable support also has come from the National Science Foundation, the Environmental Protection Agency, the National Aeronautics and Space Administration, the National Oceanic and Atmospheric Administration, and the Department of Energy.

I met Stan Ulam only once, at the D. H. Lawrence Ranch in New Mexico many years ago. His wit and insights, however, were apparent even in the brief interactions we had, and it has been an honor and a privilege to produce this book as a tribute to him. I never met Charles Darwin, the father of complexity theory in ecology, but my intellectual debt to him will be obvious. Finally, I thank my colleagues at the Santa Fe Institute, and especially Ellen Goldberg and Erica Jen, for the opportunity provided me to give the Third Ulam Lectures, from which this book emerged.

Simon Levin

FRAGILE DOMINION

1

BIODIVERSITY AND OUR LIVES

A Cautionary Tale

Mother Earth is in trouble, at least as a habitat for humanity. Pollution fouls our air and waters, stark evidence of our presence. Ozone holes gape above the poles. Population continues to expand, while fisheries crash. New diseases like AIDS and Ebola ravage the unfortunate, while old ones like tuberculosis reassert themselves, overcoming miracle antibiotics. Clearly, our environment is in need of attention.

So many problems, so many reasons! How do we even begin to deal with our dilemma? Should we turn all our energies to worrying about elevated temperatures, the rise of sea level, and the potential loss of coastal communities? Or do more immediate problems, such as malnutrition, pollution, and the spread of disease, take precedence? The thesis of this book is that, to have any hope of dealing with such a complex combination of threats to our survival, we must study the Earth as an integrated physical and biological system. By understanding what makes that system work, we will understand how it can fail, thereby finding a way to prioritize actions and maintain the Earth's ability to continue to nurture and sustain us.

The central environmental challenge of our time is embodied in the staggering losses, both recent and projected, of *biological diversity* at all levels, from the smallest organisms to charismatic large animals and towering trees. Largely through the actions of humans, *populations* of animals and plants are declining and disappearing at unprecedented rates; these losses endanger our way of life and, indeed, our very existence. *Biodiversity* loss provides immediate evidence of environmental change, and it also threatens the very structural and functional integrity of the Earth's systems, and ultimately the survival of humanity.

Undergirding the dynamic Earth—its atmosphere, its physical and chemical fabric, and its biological essence—is a prototypical *complex adaptive system* (CAS), one that we call the *biosphere.* It has, over ecological and evolutionary time, spawned increasing biological diversity, but simultaneously it has evolved patterns of arrangement and interaction of its pieces. The result is an integrated network, with characteristic flows of materials, energy, and information that exhibit regularity in dynamics over long periods of time. Understanding the essential features of the biosphere's internal organization, and what maintains it, is fundamental to developing a rational and effective strategy for preserving the environment with quality sufficient to sustain us, our children, and our children's children.

Human Stewardship of a Changing Environment: A Historical Perspective and a Glance at the Future

The second half of the twentieth century has been characterized by a heightened awareness of environmental change and by the introduction of measures to slow ecological degradation. The institution and implementation of adequate laws and responses has worked best for issues such as local and regional pollution, where the signals are clear, effects are closely related to causes, and the potential for solutions is therefore greatest. The familiar acronym *NIMBY* ("not in my back yard") expresses the principle that people can best be motivated to take action when the problems and rewards hit closest to home. The nature of the process of addressing local issues makes for tighter *feedback loops,* a key element in maintaining resiliency in any system.

Increasingly, however, we are being challenged by a new class of problems, including global climate change and biodiversity loss, in which the feedback loops are weaker and less specific. Change is slower, and signals less clear (hence the delay in recognizing them).

Cause and effect are more weakly linked, involving diffuse couplings among many elements. The lack of tight *correlations* makes the incentive for self-regulation weaker and the solutions less obvious. Recovery is delayed and eventually may become impossible because of irreversible shifts that may occur before solutions can be implemented.

A concern for biodiversity is not new. Humans have been fascinated with biological diversity since biblical times, and surely before: ancient texts display clear appreciation for the most basic ecological principles. In the Old Testament version of the Creation, primary producers of energy (plants) enter the picture on day three, within hours of the appearance of dry land and only after the first rainstorm. *Phytoplankton,* the tiny aquatic plants, are never mentioned explicitly, though in theory they could have come on the scene the day before; there was already plenty of water. The absence of sunlight would have created some difficulties, though only until day four; it did not seem to inhibit the other plants much. The text says simply, "Let the earth put forth grass, herb yielding seed, and fruit-tree bearing fruit after its kind, wherein is the seed thereof, upon the earth."[1] Thus, not only were there plants, but there were lots of different kinds. A good day's work.

Animal life followed two days later; it was bad enough that there was no sunlight for photosynthesis until day four, so it was reasonable not to have the plants contend with herbivory as well. But on day five things really got going: "Let the waters swarm with swarms of living creatures, and let fowl fly above the earth, in the open firmament of heaven." Specifically, this led right away to the creation of "the great sea-monsters, and every living creature that creepeth, wherewith the waters swarmed, after its kind, and every winged fowl after its kind." Just before nightfall, they were given instructions on how to reproduce, and by the next day there was a proliferation of biodiversity, marked by fish, fowl, cattle, and creeping things—all setting the stage for the arrival of the latecomers, humans.

The beauty of this story is not just that it includes a rudimentary understanding of a number of ecological principles (albeit an apparent lack of understanding of others), but also that it exhibits a fascination since the earliest writings with the diversity of the biological world around us. In fact, it recognizes that biodiversity was created before humans were even introduced as a part of that biodiversity. In a variety of ways, the biblical tale also addresses ecological function:

plants exist for animals to eat, and animals exist, at least in part, to serve humans. This proposes a view of evolutionary purpose that would not sit comfortably with most evolutionary biologists today, but it does clearly recognize the interconnectedness of nature.

In the Great Deluge in the time of Noah, the importance of preserving biodiversity is made clear by the construction and design of the Ark, the purpose of which was to preserve not only humans but "every living thing of all flesh." The focus clearly was on animals, and in particular on sexual species: "And of every living thing of all flesh, two of every sort shalt thou bring into the ark, to keep them alive with thee; they shall be male and female."[2] But obviously provisions were made for plants and asexual animal species as well, many of which could not have survived the Flood if its geographic dimensions were as great as suggested.

Human fascination with biodiversity, its causes and consequences, continues unabated, even as our own activities create perils no less severe than did the Deluge. Natural historians since Linnaeus, and before, have been fascinated with cataloging biodiversity and arranging it systematically. The search for generative explanations for these patterns culminated in Darwin's evolutionary theories, which provided a dynamic context for understanding not only existing patterns but also the emergence of diversity and its maintenance. Darwin's theoretical structure gave only the outlines of an explanation. Evolutionary theory today still struggles to develop understanding of the origins and maintenance of biodiversity and of the importance of processes such as *mutation*, *recombination*, and *selection*. These continually provide and reinforce new forms of genetic variability, thereby building and sustaining diversity.

Ecological studies, in the meantime, offer a complementary approach to an appreciation of biodiversity, elucidating the interdependence of species and their links to the physical world; more recently, recognition has grown of the manifold ways in which biodiversity, at many levels of organization, is essential to human life and to the maintenance of an environment consistent with human existence. Nevertheless, the connections between the ecological and evolutionary points of view leave fundamental issues unresolved. The biblical view of the Earth as an integrated whole resonates with much current thinking; the suggestion, however, that evolutionary forces have shaped the pieces to serve the whole, or to serve certain chosen components such as humans, is not consistent with what we know about *evolution*. There is a chasm here, by no means restricted to the biblical rendition, between these two differing perspectives on what has shaped the world's *biota*, its assemblage of biological organisms. To bridge this chasm, we need to understand how the complex biosphere has emerged from natural selection and other forces operating at small scales. We need to relate the *macroscopic* to the *microscopic*, and in particular we must elucidate the fundamental importance of biodiversity for the sustenance of life as we know and enjoy it, and the degree to which the evolutionary process operates to maintain that critical support system.

Disappearing Biodiversity

I have focused on biodiversity loss for two reasons: the seriousness of the problem, and the consequences for our lives and those of our children. According to the Nature Conservancy, which has done so much to preserve our natural heritage, one-third of U.S. plant and animal species are at risk of *extinction*, and hundreds of species may already have disappeared forever. There is no return from extinction, and the evolutionary history and biological resources embedded in each species are permanently lost when that species is lost. Globally, the dilemma is even worse—we are losing species at rates never before observed. Indeed, the dimensions of the threat go beyond simply loss of species. Maintaining populations of species in a few isolated outposts may nominally preserve those species, but it conceals the consequences of the disappearance of most populations of those species, and of the associated loss of diversity.[3] It is a bit like preserving a few specimens of a species in a zoo: technically, the species still exists, but it is doomed to extinction. That matters not just for the preservation of those species but also for the reduced services those species provide to humans.

According to Ian Turner, an expert on the forests of Singapore, nearly five hundred forest species have already been lost from Singapore.[4] Because Singapore is only six hundred meters from the Malay peninsula, it has virtually no *endemics*—that is, species found nowhere else. Thus, few of these local extinctions represent global species losses. Nonetheless, the loss of these organisms from Singapore has tremendous implications for the biology of Singapore; for example, nearly one hundred forest bird species have been lost in Singapore, deprived of their natural *habitat.* And as more Singapores emerge, the threats to worldwide distributions of these species increase. Species extinction is like an epidemic in reverse: loss of local populations contributes to loss of their neighbors, leading to spreading patterns of local and ultimately global extinctions. Thus, species loss is just the tip of the iceberg, reflecting but masking the far greater and perhaps more important loss of diversity *within* species.

Should we care? Surely respect for those who co-inhabit the planet with us, and for their right to exist, is a natural outgrowth of all of our religious and ethical teachings. But we also know that change is part and parcel of the dynamic history of the planet. Species go extinct, and others replace them. That is the way it has always been. The dinosaurs are no more, and there is not much that we would have been able to do about their predicament even if we had been there. What is different now, however, is the magnitude of the problem, the fact that we may have something to do with present and impending extinctions, and the reality that we ourselves are possibly among the endangered species. Our turn for extinction will also come, but we should avoid speeding its arrival. Our own existence is not independent of that of biodiversity: we rely on a wide range of services that other species provide, and their demise hastens our own.

In her recent book *Nature's Services,* the Stanford ecologist Gretchen Daily has assembled critical essays by some of the leading experts on our dependence on natural systems.[5] Her book provides the best summary available of what humanity is given by nature. At one level, these services are obvious. We cannot eat plastic, and hence we depend on plants and animals for our very nourishment, as well as for materials for building shelter and powering our machines. Less appreciated, perhaps, is the degree to which we depend on natural products for medicines. Two-fifths of all prescription drugs in the United States contain active ingredients originally derived from nature,[6] and these products represent only a small portion of what could be discovered.[7] Antibiotics, anti-cancer drugs, birth control pills, and a whole range of products derive from natural plants, which still hold many secrets for dealing with human disease and frailty. As we destroy these natural storehouses, we bury forever their potential to provide remedies to ameliorate the human condition. Just as we will never hear any of the symphonies or concerti Mozart would have composed had he not met an untimely death, so too will we never know what treasures lie in those species prematurely eradicated. What pain not to hear the gems that Wolfgang surely would have produced; how much greater the ache to see human suffering that might have been prevented had the secrets locked up in extinct species not been lost.

Less obvious than the direct services that ecological systems provide humans in the way of food, fuel, fiber, and pharmaceuticals are the indirect services we receive through the maintenance of natural *ecosystems.* Freshwater ecosystems dilute, detoxify, and sequester poisons that could otherwise cause untold damage to human and other animal populations. As Sandra Postel and Stephen Carpenter argue in *Nature's Services,* waterborne diseases are the major source of mortality among the poor of the world, primarily because those peoples lack access to safe drinking water.[8] Concern for the quality and quantity of water is paramount on any list of environmental problems that society must address. Just maintaining water supplies is not sufficient; the natural communities that live in freshwater must also be sustained to keep water supplies pure. Aquatic biodiversity is essential to human health.

Soil loss and degradation represent a global threat not far behind water on the list of endangered natural resources. The activities of humans have led to huge soil losses in the last half-century, with consequent leakages of water and nutrients, leading to the potential for declining agricultural productivity and increasing desertification.[9] In Indonesia, about one-fifth of the country has been lost to erosion[10]; the effects have been particularly devastating on the island of Java, which in the late 1980s was losing the capacity to satisfy the needs for rice of more than 10 percent of its population every year.[11] As with water, maintenance of soil as a resource also requires safeguarding its quality by preserving the natural communities of organisms in the soil.

Natural systems help secure the very conditions that permit our survival, moderating weather, stabilizing soil, coastlines, and climate; influencing our atmosphere; and in general making it possible for humans to exist and persist. They do not maintain those conditions in order to preserve the world for *Homo sapiens;* rather, *Homo sapiens* exists because those conditions permit it to do so. The subtlety of this distinction should not disguise the importance of the lesson. That is, we cannot count on the biosphere to maintain the biota and environment to our specifications; the world is constantly in transition, more so today than ever before in recorded history. Biodiversity is being lost at alarming rates, and with it the services that sustain the human population. Should we care? We had better care if we care about our own survival.

But what can we do about the problem? Aldo Leopold, in his elegant plea for rationality in dealing with biodiversity loss, said, "To keep every cog and wheel is the first precaution of intelligent tinkering."[12] But the global ecosystem is not a machine just out of warranty; rather, it has been out of warranty since Eve met the serpent in the Garden of Eden. It has survived because of functional redundancies—that is, the existence of multiple species that fill similar ecological roles. This form of diversity, a level up from diversity within species, provides the biosphere with the potential for alternative ways to maintain its functioning even in the face of changes. This *stability* is termed *homeostasis,* meaning the maintenance of state. Why the Earth enjoys those redundancies is a problem worthy of deep examination, and one that admits no easy solution. At this point, we should simply rejoice that it does benefit from them, and recognize the potential consequences of their loss. To understand the generation, maintenance, and importance of structure, of heterogeneity, and of redundancy of function is foundational to the theory of all complex adaptive systems, from individual organisms to whole economies. Hence, the incentive is strong for exploring biodiversity within the context of the analysis of such systems; indeed, this orientation provides the central theme for the rest of this book.

Dissecting Biodiversity

To say that not every cog or wheel or rivet is essential does not imply that none are. Paul and Anne Ehrlich, eloquent campaigners for common sense in preserving our environment, have provided a potent and oft-cited metaphor that emphasizes this point:

> Ecosystems, like well-made airplanes, tend to have redundant subsystems and other "design" features that permit them to continue functioning after absorbing a certain amount of abuse. A dozen rivets, or a dozen species, might never be missed. On the other hand, a thirteenth rivet popped from a wing flap, or the extinction of a key species involved in the cycling of nitrogen, could lead to a serious accident.[13]

The point is that the cumulative effects of species loss may be devastating even if the loss of individual species is not. This does not imply that every rivet is equally important, or that every species is equally valuable. Critically placed rivets may play pivotal roles in any machine, just as critical species may play pivotal roles in the functioning of ecosystems. One of the most dramatic examples is the California sea otter, the lovable creature that is the delight of tourists up and down the West Coast of the United States. The otter disappeared over much of its historical range during the nineteenth century, owing to excessive hunting for its pelts. Protected as a marine mammal, the otter has made a dramatic comeback, and a population of several thousand exists once again in California, Oregon, and parts of Washington. With the sea otter present, the coastal ecosystem is very different than it was without it. The sea otter feeds heavily on shellfish, especially sea urchins; when otters are around, the number of urchins is much lower. The chain of consequences does not stop there. The urchin feeds on the large kelp that grace the coastal waters; with fewer urchins feeding on them, kelp populations expand, buffering the shores from wave activity and increasing the supply of nutrients available to fish populations. The resurgent otter thus has had both ecological and economic consequences: in coastal fisheries once largely dedicated to shellfish, finfish populations now predominate. It is no surprise that shellfishermen are not big fans of the cuddly otter.

The California sea otter is perhaps the most impressive example of a *keystone species,* a term introduced into the ecological literature by my friend and colleague Robert Paine, one of the leading ecologists of our time. Paine has spent the last thirty-five years studying the spectacular rocky intertidal communities of the Washington coast, documenting the mechanisms that sustain the high diversity characteristic of the outer coast. The most famous and far-reaching of his discoveries was that the multicolored sea star, *Pisaster ochraceous,* plays a role in the intertidal similar to that played by the otter elsewhere; indeed, Paine's work predated studies of the otter's role, and his insights led James Estes and John Palmisano to evaluate the otter in the same light. *Pisaster* feeds preferentially on the large mussel, *Mytilus californianus,* which is the bully of the intertidal in that it can outcompete all other species that, like the mussel, attach to the rock surface. In areas of the intertidal where the starfish cannot survive, for example, in the upper zones where the lack of protection from the hot sun would lead to baked starfish, the mussel completely dominates the rock surface, eliminating all other species from that resource except in areas where it cannot survive owing to desiccation or wave damage. In the presence of the starfish, however, the picture is very different: the mussel is reduced to manageable population densities, leaving plenty of room for other species to find temporary safe haven. The starfish is the original and prototypical keystone species.

Since Paine's original and seminal work, the notion of keystone species has remained attractive and led to deeper understanding of the dynamics and functioning of a wide variety of ecological communities. But the situation is not usually that simple. More generally, one finds groups of species that perform critical tasks but whose roles in maintaining the integrity of the ecosystem are to some degree interchangeable. These groups may involve species high in the food chain—the starfish and the otter are considered top *predators* in their systems, feeding at the apex of the pyramid of energy flow, what ecologists call the *trophic pyramid.* But more generally, such groups involve those less charismatic species that form the foundation of the networks of energy transfer. These may be plants, which convert the sun's energy into *biomass* through *photosynthesis,* or microbial organisms, which transform nutrients into forms that are usable by higher organisms.

The role of microbes in the *nitrogen cycle* provides a case in point. Nitrogen is plentiful in the atmosphere; indeed, gaseous nitrogen makes up about 80 percent of the atmosphere. But nitrogen in that form cannot be used by higher organisms, though it is essential to their survival. Bacteria and cyanobacteria fix nitrogen; that is, they convert it into other compounds such as ammonium and ammonia, which can be oxidized for energy or assimilated by some organisms. Because *nitrogen fixation* is essential for the maintenance of ecosystems, a keystone group comprises the species that perform this role; they are essential in a way even more basic than are the keystone predators to the structure and functioning of the systems in which they exist. Groups of species of microbes similarly mediate the other critical stages in the nitrogen cycle: *nitrification,* which oxidizes ammonia to produce nitrates that plants can easily assimilate, and *denitrification,* which reconverts nitrate into nitrogen gases. Thus, the concept of a keystone group, or a *functional group,* is a natural extension of Paine's ideas. To understand the structure of an ecological *community* is to understand what the keystone functional groups are, and how they relate to one another.

The problem of identifying functional groups, however, is not so easy. What are the crucial structural components in ecosystems, and how much functional redundancy exists within them? If one species of nitrogen fixers were to disappear, would the loss be noticed? How many rivet species could be spared? A functional group is like a bigger version of a species. Within a functional group, there is also diversity, defined largely by the number of different species that make it up; within a species, there is diversity, defined largely by the number of different genetic types that make it up. When a *species* loses diversity, it is more *susceptible* to population crashes, and possible extinction, because it does not have the flexibility to respond to changing environmental conditions. It is the diversity or heterogeneity within a population, its genetic variability, that is the raw material for evolutionary change. The same may be said of a functional group, but the players are different. When a functional group loses diversity, it is more susceptible to collapse because it does not have the genetic diversity that provides the flexibility to respond to changing environmental conditions. Were the functional group that fixes nitrogen to collapse, so too would life as we know it; any major perturbation of the nitrogen cycle, in fact, would initiate catastrophic changes. Yet we know little about how to evaluate the diversity within the group of nitrogen fixers, or about what maintains it. Is there excess buffering capacity, maintained either by our good luck or somehow by natural selection? Or has this functional group evolved to the edge of disaster, where any major perturbation would have immediate consequences? The pity is that we do not yet know how to answer this question; it is a classic challenge in the theory of self-organizing systems.

Complex Adaptive Systems

Self-organizing systems have been the fascination of scientists from a diversity of disciplines because the concept of *self-organization* provides a unifying principle that allows us to provide order to an otherwise overwhelming array of diverse phenomena and structures. By self-organization I mean simply that not all the details, or "instructions," are specified in the *development* of a complex system. The specifics are in the often simple rules that govern how the system changes in response to past and present conditions, rather than in some goal-seeking behavior. The distinctions between these may seem subtle, and formal definitions are elusive and often debated. In general, however, self-organization characterizes the development of complex adaptive systems, in which multiple outcomes typically are possible depending on accidents of history.

Complex adaptive systems include a wide variety of examples that are familiar to us all. The nervous system of a young child learning to cope with her environment is an example of a complex adaptive system, since she builds on her experiences and changes her behavior according to some scheme of local rewards. Patterns of behavior and conceptual frameworks, termed *schemata*, emerge and help to facilitate future learning experiences, but the immediate modifications of behavior come from local *feedback*.[14] Indeed, not just the nervous system but an entire developing organism is a complex adaptive system, taking form from an initially featureless fertilized egg, without benefit of a blueprint.

To varying degrees, corporations, whole economies, ecosystems, and the biosphere represent other examples of complex adaptive systems.[15] The essential features of a CAS are

- *Diversity and individuality of components*[16]: This feature also implies that there are mechanisms, such as mutation or genetic recombination, that continually refresh diversity.
- *Localized interactions among the components:* In natural systems, these interactions include processes such as competition for food, predation, and sexual reproduction.
- *An autonomous process* (such as natural selection) that uses the outcomes of those local interactions to select a subset of those components for *replication or enhancement.*

I explore each of these features in greater detail in this book, but it should already be clear that the biota of the Earth fit this description to a tee. Indeed, natural selection, by feeding on and enhancing the diversity and individuality of the organisms that inhabit the globe, is the prototypical "autonomous process."

By viewing the Earth as a complex adaptive system, we are not just giving it a fancy new name. Identifying it with other systems that have the same organizing principles enables us to import knowledge that has been learned from research into those systems. And by treating all complex adaptive systems as a family of related entities, we can abstract essential features and explore through simplified *models* the properties that all such systems have. This is a very powerful technique, no different in philosophy from the use of models within a single discipline.

What are some of the characteristic aspects of complex adaptive systems? Most fundamental is the *heterogeneity* of the components, which provides the variability on which selection can act. Typically, through *nonlinear* interactions among those components, they become organized *hierarchically* into structural arrangements that determine and are reinforced by the *flows* and interactions among the parts. These four aspects—heterogeneity, nonlinearity, hierarchical organization, and flows—are key elements of complex adaptive systems.

In his stimulating book *Hidden Order*, John Holland discusses these properties, and their implications for how we study nature.[17] Because heterogeneity, or diversity, is the most basic feature, it is the unifying theme for this book. How should we measure and understand the importance of ecological diversity, specifically biodiversity? How is it distributed? What maintains it? What are the consequences of losing it? To address these questions, we need to organize the vast mass of data into categories, as Carl Linnaeus did for species, trying to lay bare the *intrinsic* hierarchical organization through the perspective of our own filters. Through this process, called *aggregation*, we classify distinct elements into categories, suppressing differences among them in order to emphasize their commonalities and to make clear their differences from elements in other categories.[18] The biological species is such a unit, a collection of distinct entities whose similarities of form and function encourage us to treat them as variations on the same theme. In the same way, a functional group can be viewed as such a category for the functioning of an ecosystem. Different species within a functional group have different properties, but they also fill similar roles within an ecosystem. For purposes of understanding an ecosystem, we may group them together in the same way that biologists group distinct organisms into a population, or distinct populations into a species. These functional groups, aggregates of individual agents, Holland calls *meta-agents.*

The other key aspects of complex adaptive systems, in Holland's lexicon, are also well illustrated in ecosystems. Nonlinearity refers to the fact that effect and cause are disproportionate, so that small changes in critical variables, such as the numbers of nitrogen fixers, can lead to disproportionate, perhaps irreversible, changes in system properties. Changes in environmental conditions or exploitation patterns can trigger qualitative and largely irreversible changes, such as desertification, in ecosystems. Not surprisingly, ecosystems show the same diversity of patterns and behaviors, and the same dependence on historical accidents, as is seen in other complex adaptive systems. Brian Arthur has emphasized the importance of similar accidents in economic systems, a phenomenon that economists refer to as *path dependency.*[19] The market success of introduced products, and indeed of political and economic philosophies, depends on their being introduced at just the right time, and under just the right circumstances. Path dependency implies, among other things, that the time and circumstances for particular ideas may never come, except perhaps in one of the infinity of parallel fictional universes in which all possible realizations of the world's evolution are played out.

There is no unique way to describe an ecosystem, any more than there is a unique way to describe an economy or a nation. Meta-agents are aggregates of agents and of smaller meta-agents, and themselves may be bundled into even larger mega-meta-agents. Any system is a mess of overlapping hierarchies of aggregations, limited in any particular description only for the convenience of the observer. For any such simplification of a system's overwhelming complexity, however, there will be flows among meta-agents, as well as flows within. In ecosystems, flow (Holland's final basic property of complex adaptive systems) refers to the flow of nutrients, the flow of water, the flow of toxicants, the flow of energy, the flow of individuals, and the flow of information.

It may well be that natural systems are not so very fragile; they are, after all, complex adaptive systems that will probably change and become new systems in the face of environmental stresses. What is fragile, however, is the maintenance of the services on which humans depend. There is no reason to expect systems to be robust in protecting those services—recall that they permit our survival but do not exist by virtue of permitting it, and so we need to ask how fragile nature's services are, not just how fragile nature is. These questions are perhaps the fundamental ones in the ecological sciences today, and they will occupy our attention for the remainder of this book.

To manage the Earth's systems and ensure our survival, we have to harness the natural forces that organize the biosphere rather than fruitlessly try to resist them. The biosphere is a complex adaptive system whose essential structure has emerged in large part from adaptive changes that were mediated at local levels rather than at the level of the whole system. Humanity's program must therefore be to understand those changes, the forces that have shaped them, and their consequences at the larger level, and then to put that knowledge to work in determining where the pressure points are for effecting changes that will preserve critical ecosystem services. This also provides my agenda for this book. In the following chapters, I develop the notion of the biosphere as a complex adaptive system and explore its emergent nature, returning in the final chapter to the lessons that may be learned for managing our environment.

2

THE NATURE OF ENVIRONMENT

Humans, and indeed most life as we know it, can survive only under a narrow range of environmental conditions, a range that fortunately exists rather stably on Earth. But is the word *fortunately* appropriately applied here? Is the juxtaposition of life and exactly the environment needed for its survival a bizarre coincidence? The result of grand design? Or is it the outcome of a coevolutionary process that has shaped life forms and environment to fit one another?

This chapter explores these issues, building from the premise that the biosphere is a self-organizing system in which global patterns emerge from evolutionary innovations that arise and spread locally. In this view, selection acts most rapidly and most forcefully at small scales, where feedback loops are tight, and much more weakly and slowly at broader scales, where feedbacks are diffuse. This applies especially to the evolution of human behaviors; what often passes for altruistic or group-oriented actions can generally be traced to hidden benefits that reward the *altruist,* or its *genes,* on fairly short time scales. Therein lies a lesson for environmental management: people behave most responsibly when they can most clearly see a payoff. This may seem a cynical and pessimistic view of human nature, but I believe it is also a realistic perspective that can help us better achieve sound environmental policies. We must provide incentives that directly or indirectly encourage individuals and corporations to be global citizens. An evolutionary perspective thus is essential not only for understanding our world but also for learning how to manage it.

Adaptation and Design

The essence of evolution, and more generally of complex adaptive systems, is that chance and choice, given enough time, make a powerful combination for change. This was Darwin's revolutionary thesis that turned popular thinking about the purpose and necessity of our world on its head. There were no guiding principles from high above. Chance generated *variation;* choice winnowed that variation. I do not use the term "choice" in the sense of a sentient being making rational decisions but rather in the sense of its synonym, "selection." Darwin termed this process *natural* selection to distinguish it from the *artificial* selection of the breeders. But the same process is at work across a huge range of systems, from organisms learning to cope with their surroundings, to the hierarchical development of corporations, societies, and cultures, to the very formation and evolution of the galaxies. These are not examples of grand design but rather of the unfolding of the consequences of chance and choice: the continual generation and exploration of randomly generated innovations, and the reinforcement of some at the expense of others.

This is not to say that design is unimportant. Today has been a good day to stay home and start writing a book. Despite the chill outside, I feel quite comfortable sitting at my word processor. It is 68 degrees Fahrenheit in my house, at least at the site of the thermometer. It has been 68 degrees all day, and it will stay 68 degrees until my wife or I decides it should be otherwise. The somewhat antiquated thermostat in the dining room still works effectively in providing the feedback and control mechanism that regulates the house's internal temperature; it has been designed to do so and will continue to do so until it breaks down.

I feel comfortable not just because of the ambient temperature but also because I have no fever. It is 98.6 degrees in my body, and it will stay at that level until my own internal thermostat breaks down. The even more antiquated thermostat that controls my body's temperature still works effectively in regulating it, deviating only slightly when I am ill. It has not been designed to do so in the same sense that an engineer has designed the household thermostat; rather, my internal thermostat has evolved under the influence of natural selection because it proved to be better than alternative mechanisms for maintaining the *fitness* of its organism. It was not necessarily better than all conceivable mechanisms, simply better than those that were reasonably similar to what has evolved.

Natural selection is a local mechanism that relies on the constant generation of slight modifications of existing patterns and then chooses among them based on what works best in getting genes into the next generation's gene pool. It tinkers. And its work is never done, since the physical and biological environment is continually changing. Though it has no global perspective, selection works remarkably well in finding solutions to the never-ending problems confronted in what Lawrence Slobodkin has termed an "existentialist" game against nature.[1] Extending Slobodkin's imagery, playing the evolutionary game is like playing a pinball machine in the local diner: the payoff for evolutionary success is simply being allowed to continue playing the game.

It was the great biologist François Jacob, in a wonderful and instructive essay in *Science* magazine, who introduced the notion of tinkering as an analogy to how evolution works.[2] Evolution does not proceed by grand design but acts simply by modifying past forms to produce small improvements. Engineers, of course, also tinker. They are not limited, however, to tinkering, and new ideas can lead to breakthroughs that may be very different in concept than their predecessors. Still, engineering innovations may also proliferate slowly, constrained by history, custom, and simply the time and cost that it takes to spread and displace established technologies. The abacus remains the favored tool of calculation in many cultures. Even among my own university colleagues, it is striking how wedded to pencil and pen and typewriter we of the senior generation were; we long resisted turning our lives over to the computers that seem to have been grafted onto our children at birth. (I am now, however, an unregenerate word-processing proselyte.) Thus, the distinctions between what engineers do and what natural selection does may not be as crisp as they at first appear. Engineering innovation does involve a fair amount of tinkering, although in principle there are no limitations on genuinely novel solutions.

Mutation, Recombination, and Sex

The difference between evolution by natural selection and change by grand design hence may be more one of degree than of absolute difference. The fundamental source of novelty in evolution is mutation, and the basic process involves a mutational change of a single gene. This would appear to make evolution a very conservative process, because organisms that disagree in a single gene out of many millions might not be expected to function very differently. Indeed, single gene mutations usually produce variants that are not too different from the existing organisms, but there are exceptions. Some are familiar to us because of their harmful consequences: genes that cause genetic defects like cystic fibrosis, for example. Such very harmful mutations, however, are not drivers of major evolutionary change: because of the disadvantage they impose on their *hosts,* they are quickly eliminated or kept at low frequencies as recessives.

Single gene mutations, however, can also confer a selective advantage on an organism and may, under some circumstances, rapidly spread. A mutation in a *pathogen,* for example, although it may be harmful to the host organism, can proliferate by helping the pathogen overcome host defenses. Bacterial *resistance* to an antibiotic may be coded in a single gene, and a strongly selective environment (the use of the antibiotic) finds and rewards such genes as soon as they make their appearance. We do not need to be reminded of how quickly diseases such as tuberculosis have developed resistance to our treatments because we have overused antibiotics. Although it may also occur in nature under rather weak selective regimes, bacterial resistance more generally evolves in acute selective environments—usually orchestrated by humans—where the intensive use of antibiotics to treat infections draws sharp life-and-death distinctions. Under such conditions, a single gene that conveys any advantage is strongly favored. The appearance and spread of *traits* controlled by single genes also occurs in other highly selective environments; examples include plants' evolved tolerance to heavy metals and resistance to pathogens (in particular where plant breeders have helped the process along by selectively breeding for resistance) and the reciprocal *adaptations* by which pathogens overcome those single gene defenses.

More generally, traits of ecological importance are likely to involve the interaction of more than one gene. Beneficial mutations that involve several genes are, of course, much less likely to arise, but evolution has had plenty of time, long enough for such coincidences to happen. As I discuss below, the evolution of multigene traits can be facilitated in sexual organisms. Through the magic of sex, possibly harmful mutations that would be immediately eliminated in asexual organisms may be united to produce genetic combinations that convey a distinct advantage to their bearers.

When a favorable mutation arises in an asexual organism such as a bacterium, there may be no stopping it. As the bacterium divides and makes copies of itself, it spreads its new idea rapidly. Sexual organisms, on the other hand, must find consenting partners before their genes can be reproduced. Even then, there is the likelihood that the innovation will be *recessive;* that is, it will have no effect on the fitness of individuals who receive the gene only from one parent. The dynamics of such recessive traits, when they are rare, are largely random; as such, there is a substantial chance that they will be lost in the sexual lottery (by which offspring inherit genes from their parents) and never reach the levels needed for establishment in the population.

Sex is not entirely a bad idea, however, as already suggested above. Bacteria use a form of sex to transfer genes from one mating type to another by conjugating and exchanging pieces of *DNA,* called *plasmids,* that are separate from the bacterial *chromosomes.* Thus, plasmids are spread like infectious agents, though they typically carry benefits for their bacterial hosts rather than harming them. Antibiotic resistance, for example, usually is carried on plasmids and hence can be spread much more rapidly through a population than if it were on chromosomes. The reason is that the essentially infectious transmission—more commonly called *horizontal transfer*—from individual to unrelated individual obviously speeds the rates of spread. And this is not an accident. Imagine that a mutation conveying antibiotic resistance arose in one organism on a plasmid, and in another on a chromosome. The mutation on the plasmid proliferates more readily because it is not restricted to the daughter cells, or even to a single host species. Consequently, it is much more likely that when we observe antibiotic resistance, it is to be found on a plasmid. Hence, this form of sex is good not only for the plasmids but also for the bacteria, because it speeds their ability to adapt rapidly. Of course, bacterial sex is not such a good thing for the host organisms, but that is none of their business.

The case of plasmid transfer of traits illustrates that, when favorable mutations arise, selection also can operate to facilitate their transmission and make the process of spread faster. More generally, this demonstrates that evolution does not take place just within given rule systems; evolution itself may act to modify the rules of the game, changing the board on which the evolutionary contest is played. The *coevolution* of organisms and environment, and the development of the rules of action and interaction, represent the essence of the challenge in understanding the structure and functioning of ecological systems.

More conventional sexual practices, such as those between consenting animals, also carry potential benefits. Sex leads to recombination, that is, to the shuffling and reassortment of gene combinations, thereby generating the variety on which selection may then operate. Furthermore, as suggested earlier, it provides a way to combine and recombine innovations that alone would be harmful to the organisms that carry them. Mutations that can potentially work together to improve the fitness of an organism may arise independently in different organisms; because the genes require each other to be effective, they convey no advantage to their bearers when they arise independently and may even be disadvantageous. But sex can bring together these two new genes in a single individual, the offspring of the mating, and allow the joint benefits to be expressed.

A closely related example is the gene that conveys resistance to malaria and hence has been favored by natural selection in populations where malaria exposure is high. The difficulty is that, in double doses (which occur when the same gene is inherited from both parents), this gene causes blood protein hemoglobin malformation, leading to anemia due to sickling of cells or other causes. On the other hand, the alternative form of this gene (the type that is common in malaria-free areas) provides protection against anemia, but no protection against malaria. The most fit individuals are those who receive one form of the gene from the father and the other from the mother. Again, sex makes possible the mixing of characteristics that are beneficial only in combination.

The evolution of sexual reproduction itself thus provides another example of how the rules of the game, which reflect higher-order processes than simply selection for a particular trait, are themselves modified through the cumulative selection pressures on many traits. In the same way, the combined weight of multiple small-scale processes can accumulate to help shape other patterns of interaction, and hence the structure and function of ecosystems, from small scales to the biosphere. Natural selection, together with other drivers of evolutionary change such as mutation, recombination, environmental factors, and simply chance events, provides the central organizing principle for understanding how the biosphere came to be, and how it continues to change. No teleological principles are at work at the level of the whole system, or even at the local level. The biosphere is a complex adaptive system in which the never-ending generation of local variation creates an environment of continual exploration, selection, and replacement.

One may be tempted to think of this process as local repair, but that would be misleading. There is no guarantee that local changes will be good for the system as a whole, nor should they be necessarily. Mutation and recombination, for example, regularly spin off new strains of the influenza virus (or old strains ready to make comebacks). Natural selection does not take into account that the spread of these strains will cause global suffering; nor does it attempt to examine whether such suffering will be good or bad for the biosphere. The concept has no meaning for the evolutionary process. Novel strains have an advantage over resident ones in having a larger number of susceptible hosts to exploit, and this provides them a selective advantage that may lead to their spread. The same forces are at work in the spread of any new or emerging disease and shape the global epidemiological environment. What is obvious in the case of emerging or reemerging diseases is equally true for the spread of any novelties. Local advantage is necessary and sufficient for that spread; global feedbacks, though they occur, are much weaker and act over much longer time scales.

Evolution and the Biosphere

The previous section provides what at first glance seems a somewhat chaotic view of biospheric evolution. Local accidents—vagaries of history that are frozen and resist displacement—become woven into the tapestry. Is the world really so disordered, or are there regularities and stabilizing feedbacks that emerge? No observer of the physical and biological Earth can fail to be impressed with its regularities and with the fact that it functions as a system. Indeed, to some the system is so buffered against change that they liken it to a superorganism, which has governed the evolution of its parts in order to ensure its own survival. This philosophy is embodied in a concept known as *Gaia*.

Gaia is a hypothesis first put forward by the geologist James Lovelock and developed further by Lovelock with the biologist Lynn Margulis in the early 1970s to explain the maintenance of stability in the properties of the biosphere, especially in those properties that are essential to sustaining human life.[3] Gaia, the thesis of two brilliant scientists wrestling with some real scientific puzzles, has, however, become transformed into a virtual cult religion because of its implications that we all have a common destiny, united in the interests of great Mother Earth. The philosopher David Abram writes:

> When we consider the biosphere not as a machine, but as an animate, self-sustaining entity, then it becomes apparent that everything we see, everything we hear, every experience of smelling and tasting and touching is informing our bodies regarding the internal state of this other, vaster physiology—the biosphere itself.[4]

This is a long way from the original ideas of Lovelock, whose principal claim was simply that there were feedbacks from the biota that served to regulate the atmosphere.[5] That claim is quite consistent with prevailing ecological theory,[6] and indeed it is the basis of concerns about such phenomena as the disappearance of tropical rain forests. A stronger form of the Gaian claim, however, is that the biota and the atmosphere have coevolved such that the biota regulates the atmosphere to just the conditions needed for its survival.

At the simplest level, there must be an element of truth to this claim, although what is termed evolution means very different things for the biota and for the physical world.[7] However, the corollary that some would draw, that the Earth has self-correcting mechanisms that will compensate for any insults to it, is patently false and dangerous, for it encourages a reckless attitude toward the environment.[8] Nor is such a view to be found in the writings of Lovelock or Margulis, whose principal argument has been simply that the study of the Earth must rely as much on biology as on the physical sciences. The latter point is unassailable; it is only the egregious exaggerations and distortions that have led to strong resistance to the concept in much of the scientific community. At its core, the claim that "the Gaia hypothesis forces us to consider the cumulative, that is, global, effects of . . . local phenomena" resonates with much that I have written in this chapter.[9] There are, unquestionably, feedback loops—and in particular homeostatic mechanisms—that maintain the Earth's environment within fairly narrow bounds. The question is, how did those homeostatic mechanisms arise? And has the biota itself had anything to do with the development of those mechanisms or is it simply the beneficiary?

The truth lies somewhere in between the extreme views. Innovation and evolution, as already described, occur first at local levels and spread from there. As new genetic types proliferate, they influence their environments, though that influence becomes more diffuse as the spatial or organizational scale increases. As influences become propagated up from the smallest scale to larger scales, feedbacks are generated down from the large scale to the small scale; these feedbacks may either constrain or accelerate local changes.

Understanding the integrated evolution of a whole system, any system, requires explication of the interplay among changes at a spectrum of scales, from the smallest to the largest. This is the essence of the study of all complex adaptive systems, and the relevance of this interplay to biospheric evolution is indeed what this book is all about.

The Importance of Scale

Evolution is the basic force governing biotic change over geologic time scales, via natural selection and related mechanisms. Central to the concept of adaptation by natural selection is variation; in the biosphere, much of that variation is expressed spatially and temporally. Variation presents itself to us everywhere we look. For example, comfortable as I am indoors and at my word processor (I am, indeed, making the transition from pen and pencil and never much liked the typewriter), I would be less comfortable were I to venture outside. But things will get better. It is 8 degrees Fahrenheit outside my kitchen window, at 8:00 A.M. on a cold January day in Princeton. It will be at least ten degrees warmer later today and is expected to warm considerably by later this week. By summer the temperature will be 95 degrees on many days; it is already 95 degrees in Bangkok, and even hotter in other places. Minneapolis, on the other hand, is eighteen degrees colder than Princeton, and it is surely not the coldest spot on the Earth. Obviously, whatever homeostatic mechanisms might operate on the temperature of the globe do not exercise very tight control at the local level. Global mean temperatures average about 60 degrees Fahrenheit over the year, though there is considerable geographic variation (and a general decline with increasing altitude). The variation is so great that the notion of global mean temperature has very little significance for most of us, and the fact that it varies relatively little over the year simply reflects the fact that when some places are hot, others are cold. For purposes of most human activities, we focus more on regions and climates, which average over many of the highest-frequency space and time fluctuations. That is, regions and regional climates show less variation than do smaller areas and daily temperatures, although they are more variable than globally averaged phenomena would be.

The central point is that variation is typically greater at small scales than it is at large scales. This is true not only for broadening spatial scales but for other ways of lumping as well. For example, stock market averages, such as the Dow Jones Index, may fluctuate more than we would like, but individual stocks show much greater variation. Indeed, therein lies the key to hedging one's bets in an even moderately conservative investment scheme. The same applies to ecological organization: fluctuations in the presence of key functional groups, for example, since they average over the fluctuations of individual species, are less than those for most of those species. Maintaining biodiversity, from the human perspective, is analogous to employing a conservative investment strategy: it increases the chance that we will not lose our entire portfolio.

Temporal averaging similarly leads to more stable dynamics, filtering out what is often called high-frequency noise. We are all aware of predictions that, at the global level, mean temperatures seem to be rising at the rate of as much as a few degrees per century. We perhaps could be forgiven for wondering why we should be concerned with such a slight change when we are used to larger changes over the course of only a few hours. Here, filtering cuts both ways. Typically, high-frequency oscillations are averaged out by organisms, which can survive brief periods of unfavorable weather or other conditions. All plants must manage the fluctuations of daily cycles, shutting down photosynthesis overnight. This degree of fluctuation has been part of the evolutionary background for the work of natural selection. Looking at longer time scales, we know that early-blooming trees can survive unseasonably late frosts. If those frosts occur every year, however, the cumulative effect can begin to take its toll, and the tree may die. Thus, a secular change in mean temperature can have a great effect on the survival of species because it presents challenges that are outside their evolutionary experiences and it happens too rapidly to allow them to adapt. Thus, there are indeed problems that would be associated with temperature changes of only a few degrees, even if uniformly applied to the globe. Sea levels would rise, threatening coastal communities. Agricultural regimes would shift, creating economic burdens. And forest composition would change, albeit less catastrophically because forests are more diverse.

As serious as the consequences of such climatic shifts might be, scientists have far deeper concerns. A global average increase of a few degrees reflects much more dramatic increases in some locations and little change in others. Furthermore, a yearly increase of a few degrees averages over periods of time when it may be much hotter and others when it may be colder. A much higher frequency of storms could be associated with this heightened variation over space and time, and it is such changes that will have the most dramatic effects on our lives.

Environmental variation over time is nothing new. In the coldest of winters, we still hold to the expectation that spring and summer will come again. Wiltshire folklore reminds us:

> *There is no debt so surely met*
> *As wet to dry and dry to wet*[10]

Many people have believed that, over the long run, good years and bad years come in cycles. The biblical Joseph built a career on successful climate prediction, interpreting Pharaoh's dreams to anticipate a cycle of plenty and famine. There are certainly ups and downs in the weather, but we are just beginning to understand the dynamics of multiyear cycles. Could Joseph have been the first real expert on El Niño, long before it got its name?

Homeostasis and Gaia

The fact that temperature fluctuates but nonetheless stays within bounds suggests some degree of regulation. Every schoolchild knows that temperature fluctuations are to some extent driven by the Earth's annual revolutions about the sun and influenced by its daily rotation about its axis and the tilt of that axis. These gyrations affect solar and terrestrial radiation and control the net energy balance. That is quite a different concept, however, from the kind of homeostatic feedback that controls body temperature or household temperature. The concept of Gaia recognizes the role that biological systems play in affecting the physical environment and suggests that these have coevolved over time to form a complex, self-regulating system.

The view advanced by Lovelock and Margulis[11]—that the biosphere has control systems that maintain the Earth in homeostasis—is a view with which most scientists would agree, within limits. Yet even if such atmospheric and climatic homeostasis exists, it has not occurred through a Darwinian process operating at the level of whole systems. Any homeostasis that exists in the Earth's climatic regime is, as we have already seen, most inexact and diffuse. Unlike the thermostat in my house, it has not been designed for control; nor has it evolved for control because of the fitness it confers on the global system. Natural selection operates by choosing among a number of different alternatives, rewarding those lines of descent with the highest fitness, and obliterating the weak and less fit. But, as Ehrlich writes, "our Earth's evolutionary past is not littered with the remains of billions of other planets where life did not manage to coevolve successfully with its physical environment."[12]

Evolutionary principles do apply to the short-term changes in the biota, more so perhaps than many would admit. Richard Lewontin, the Harvard geneticist, draws a distinction between evolution via natural selection and transformational evolution, such as describes the evolution of the universe as a whole.[13] Transformational evolution describes a unique event, the development through time of a single entity. The term could be applied to the assembly of a car at a factory by auto workers, or to a system that is self-organizing through the adaptive evolution of its parts. In the latter case, which is characteristic of any complex adaptive system, transformational evolution takes place at a macroscopic level because of adaptation—essentially natural selection—at a lower level. Surely the emergence of these system properties may feed back to influence selection at lower levels, and it can be argued that systems that have lasted for long periods of time are systems in which these feedbacks have not been destructive. Putting aside the issue of what "long periods of time" means, I still conclude that the existence of atmospheric conditions favorable to life is to a large extent a happenstance—indeed, one that is inseparable from something cosmologists call the *anthropic principle.*

The Anthropic Principle and Gaia

The anthropic principle is rooted in the observation that the conditions for life are likely to be present for a long time in only a small number of worlds. But given that such conditions do exist, life may evolve in those places. Where life is, life is. If life brings with it the ability to perceive and to question, then those who perceive and question will happen to be doing that in those few places that permit their existence. Small wonder, then, that what we the observers and questioners see around us, to our bewilderment, seems to defy rational explanation other than through divine creation or Gaian coevolution. But there is no paradox here; the presence of the conditions for life exactly where we find ourselves is almost a tautology. As Paul Davies says:

> If the existence of life depends on certain natural conditions being fulfilled, then we would not expect to observe a universe very different from the one we inhabit; put simply, the world we live in is the world we can live in. This is not an explanation of the natural features around us, only the observation that intelligent creatures could not be here to speculate about it if things were very different.[14]

Stronger, and more controversial, forms of the anthropic principle do suggest a form of system-level natural selection. Imagine, for example, that Gaian feedbacks, by providing stability, operate in such a way that the conditions favoring the existence of life happen also to be those that favor the spawning of new daughter universes with the same properties; natural selection then could operate at the level of universes. This idea is intriguing but is really only pure speculation. It sheds little if any light on how the feedbacks between the biota and the physical environment might have coevolved to provide stability, and it has no empirical support.

The Emergence of Homeostasis

The principle of homeostatic feedback control is a common one in the design of machines or guidance systems. The cruise control system on a modern automobile acts as a governor, maintaining speed at a desired level; an airplane's guidance system can make adjustments to keep the craft on course; a fuse melts when overloaded in order to protect an electric circuit. All of these are features that have been engineered to respond to the state of a system and to help maintain some desired behavior. The study of such control processes, whether in electronic, mechanical, or biological systems, comes under the general heading of *cybernetics.*

Biological systems rely on a dazzling array of cybernetic or homeostatic mechanisms. These are not the work of a master designer, however; rather, as suggested by François Jacob, they are the result of tinkering. Natural selection builds on past innovations, adopting those mutations that provide improved fitness. In this way, an endless process of change replaces one solution with another, each somewhat better suited to the ever-changing environment than were its immediate predecessors. Such evolutionary sequences are played out over and over again across space and time, each realization taking its own unique form. There is no purpose or design to this process. It is idiosyncratic and profoundly influenced by history. Where an engineer's design is involved, the very best solutions can dominate, and most devices meant for a particular purpose share similar design features. (An exception that I have never quite understood seems to be flush toilets, which take all shapes and forms in different nations. Perhaps this is because no particular design seems to work especially well, but it probably is also the result of an idiosyncratic process largely guided by historical accident.)

For the solutions that have been shaped by natural selection, the influence of history and local variation clearly enhances variety. Thus, the solutions to biological control problems appear in stunning diversity and provide an almost inexhaustible library of spectacular complexity. Homeostatic controls have evolved for a multiplicity of physiological processes, each critical to the survival of the organism. Postural feedback control allows an individual to walk. We have all marveled, for example, while watching a young child learn to detect loss of balance and to translate that information more and more efficiently into instructions to muscle systems that prevent her from falling down. And the *vertebrate* immune system is a masterpiece of feedback control, sending scouts to patrol for invaders into the body and then instituting processes to control and eliminate those invaders.

For the biosphere—that is, for the whole Earth, including its biological parts—homeostasis is at best an observation of stability under historical patterns of variation and a hope for similar stability in the future, when conditions may be quite different. Unfortunately, it is not a blueprint for survival. There are certainly radiant heat feedbacks in the climate system, for example, and these are controlled to a large extent by how the atmosphere operates to maintain energy balance at levels that allow life as we know it to exist. During nearly the first half of the history of the planet, the Earth was almost completely covered by water,[15] and today water covers more than 60 percent of the Earth's surface.[16] All that water was necessary for the development and early evolution of life, and water remains necessary for maintaining favorable conditions today. Was this a fortunate accident? The Gaia hypothesis argues that it cannot be, and that the biota must have played a role in maintaining exactly the conditions that it needed. But how could this have happened? There are no principles of natural selection operating at the scale of the whole globe,[17] and it remains unproved that evolution proceeding at the level of individual organisms and groups of organisms could give rise to the evolution of characteristics that preserve and protect the global enterprise. Would that things were that simple, and that we could relax in the conviction that humans pursuing their own selfish interests were perforce operating in the greater good of humanity. That, unfortunately, is not so. Natural selection operates at the level of what is good for the individual rather than what is good for the group—much less the whole Earth—and it is noteworthy that Darwin himself, as well as those who continue to study evolutionary processes, puzzled about why altruism exists at all, rather than about why it is not more widespread.

Such hopeful hypotheses are unnecessary, however, at least for understanding how the physical properties of the Earth came to be just what we need for survival; the anthropic principle provides a ready explanation. Indeed, the conditions that favor life are very unlikely, on any given planet, but it is not unlikely that one or a few of the multitudes of heavenly bodies would possess exactly the right conditions—indeed, exactly the necessary feedback mechanisms to maintain the conditions that would allow a stable favorable climate for more than four billion years. Under such conditions, life would be expected to arise and evolve, and the products of that evolution would be around to observe it and wonder about it. To state, "How fortunate that exactly the conditions that favor my existence happen to have arisen on Earth, where I am," is like saying, "Of all the people who have ever lived, how unlikely that I turned out to be me." In each case, the statement reflects a biased observation rather than a random selection from a universe of possibilities. Neither question could have been asked were the conditions for existence not satisfied. This paradox does not mean that there has not been some form of coevolution; clearly, there has been. It simply means that such coevolution is not essential to explain what seems like our highly unlikely good fortune. It also does not mean that humans do not influence the global environment; they clearly do—and indeed have made substantial strides in screwing it up.

Lovelock and Watson recognized the challenge of demonstrating how natural selection might operate to guide the coevolution of organisms and environment.[18] Their ingenious model, *Daisyworld,* explores evolution by natural selection among two types of daisies: white daisies, which are cooler than the local environment, and black ones, which warm it. The local warming enhances growth and reproduction, favoring the spread of black daisies. Eventually, the black daisies take over, maintaining temperature at stable levels.

Daisyworld illustrates two principles that are central to this book: the emergence of macroscopic patterns from local selection, and the importance of the biota in mediating climatic conditions. Even in its more sophisticated later versions, however, Daisyworld is at best a caricature of how selection might operate to enhance homeostasis.[19]

Evolution Above the Species Level

Despite the uncertainties about Gaian coevolution, it remains obvious that evolutionary processes, including selection, have been important in shaping our environment. This applies not only to the natural environment but to our social systems, our economic systems, and our cultures. The physical environment has provided a context in which evolution can take place, but the emergence and evolution of life has been a self-organizing process that ultimately must feed back to influence the total environment at multiple scales. From the myopic selection of properties that benefit individuals, populations of similar organisms arise. From the interactions of these populations with other populations, species emerge. From the interactions of species, mutualisms such as pollination provide feedbacks that affect the further evolution of the partners. Individuals and populations compete for resources and often find ways to share them to permit *coexistence*. Coexistence is not a given, however; the exclusion of inferior *competitors,* at least in some habitats, is common. Finally, exploitative relationships such as predation and parasitism establish conduits for energy to flow from the sun through many layers of users. Through the complex interplay of all of these mechanisms, ecological communities take shape as loosely defined assemblages of role players, not as superorganisms with a common purpose and destiny.

Is this story so different from how human societies grow and are maintained? Humans think first about themselves, and about maximizing their benefits through their actions. This extends quite naturally to our children, but that is the product of natural selection working to increase the representation of our genes in future generations. Cousins and more distant relatives benefit from our largesse, but to an extent that decreases as the closeness of our relationship decreases. Of course, our tendency to help our relatives, in order to promote our common genes, leads to a tendency to favor those who look like us, since they are more likely to carry similar genes; thus, it also leads to racism and tribal conflict.

Nonetheless, altruism does arise in human societies, typically to benefit kin or as reinforcement for the establishment of temporary coalitions. Called reciprocal altruism by behaviorists, this is not really altruism, of course; it is enlightened self-interest: both participants in a coalition recognize that there is more to gain from a partnership than from competition. Individuals and coalitions compete for resources with other individuals and coalitions and find ways to share them to permit coexistence. Exploitative relationships develop, allowing resources to flow from workers to venture capitalists and other investors through a complex web of interactions. Human communities and societies take shape, self-organizing from selfish interactions initiated at the level of individuals, within sets of rules that are imposed by the laws of individual societies. Indeed, the laws themselves coevolve with societies, complicating and enriching the analysis further. To understand how societies work, one must understand how individual behaviors evolve and coevolve to allow the development and maintenance of societies, and the degree to which this sets up structures that resist change that might be in the common good. To improve societies—and to preserve the fragile biota that allows human societies to thrive—we must understand the strength of the selfish impulse, the importance of the individual's actions in governing his or her behavior.

How do we get from A to B when B is better for everyone, or for almost everyone, but when no individual acting alone can initiate change without incurring personal costs sufficient to inhibit enlightened behavior? Not an easy question; if it were, we could eliminate wars and reduce crime in the streets. The famous *game theory* example, Prisoner's Dilemma, makes the problem clear. In the game of Prisoner's Dilemma, two prisoners are interrogated separately about their parts in a crime. Knowing that their best strategy is to hold to their common alibi, which will prevent their incarceration and let them off with light fines, each is nonetheless tempted to turn state's evidence and get off scot-free, allowing the accomplice to pay for the crime with a very harsh penalty; this is a strategy that works only when there is just one rat. When both relent, however, and spill the beans, both end up with long prison terms. They would have been far better off to cooperate, but cooperation is a difficult strategy to maintain in this case without mutual reinforcement. Separating the prisoners has reduced the potential for that reinforcement, the stability that comes with mutually assured destruction, and has led to common defection.

A second example, introduced to me by my postdoctoral associate Gregg Hartvigsen, is perhaps even more instructive. Using a game invented by the economic theorist Martin Shubik, Gregg offers his students the opportunity to bid on a dollar; high bid wins, but the two highest bidders must pay. The auction soon reaches a stage where the actual payoff is irrelevant. A bidder who has offered $.95 may be tempted to go to $1.05 if someone else has bid $1, even though the prize is not worth $1.05, because the $.95 must be paid anyway. Otherwise, if the auction were allowed to end, the top bidder would pay $1 to gain $1, walking away with nothing, while the second-highest bidder would be out $.95. In fact, the student is bidding only $.10 more than is already obligated, in the hope that that bid will capture the prize. But once the $1.05 bid is in, the $1 bidder will have the same incentive to escalate, and prices can continue to spiral upward. The problem with this game is that there is no stopping it, and the bidding can go to several dollars or more. The auction soon reaches a stage where the actual payoff is irrelevant compared to the price that must be paid. Again, cooperation would have been the best strategy, but by the rules of the game, cooperation is not permitted. If the game is played repeatedly, individuals may develop strategies that mimic cooperation, just as repeated playing of *Prisoner's Dilemma* may lead to forms of cooperation. Gregg typically does not perform the experiment of playing the game multiple times, however, because the students would quickly catch on to the fact that he does not make them pay up.

The Evolution of Altruism

The simple game-theory examples of the last section carry profound implications for the evolutionary theater. The evolution of altruistic behaviors can be understood only in terms of payoffs to individuals, or to the genes that control behaviors; that is, altruism evolves most easily when selfishness is greatest. The simplest such payoff scheme is reciprocal altruism, or mutual back-scratching: short-term altruistic behaviors are held together by expectations of similar payoffs on rapid time scales. Scholars read each other's papers and review each other's grant proposals because they want their own papers read and grant proposals reviewed. These are simple cases of coalitions: individuals binding together, with common rules of membership, in order to ensure some mutual level of reciprocal altruism. Communities and cultures arise, laws and customs emerge. Yet the more local the laws, the more likely people are to feel committed to the spirit as well as to the letter, because the more likely they are to see benefits to themselves. The United Nations has not really engaged the loyalties of any peoples because its possible benefits would need to be shared with so many; paradoxically, this observation explains why such global bodies struggle to survive and succeed, except in the science fiction context of interstellar travel and galactic wars. In such fantasies, the nations of the world, faced with the common threat of invasion by aliens, band together and international coalitions thrive, at least in some of the most optimistic scenarios of "star wars." What once was global has become local, and the *positive feedback* loops become tightened. Altruism seems to exist because of a conflict between factors at different scales: the suppression of local differences to combat some greater threat. It would be nice to know that altruism could arise without the common enemy, but it is hard to think of examples where that is clearly true. Even in the simple examples that arise in the plant and microbial worlds, the coalitions that masquerade as mutualisms achieve, through their interaction, some competitive advantage over plants and microbes that do not form coalitions.

Coalitions may be held together in various ways. Neighborhood action groups provide examples in which geography alone does the trick, and cities and nations are elaborations of this theme. Labor unions and credit unions are examples where common interest provides the glue, and the common purpose. Genetics provides perhaps the strongest link, often reinforcing geographical or other factors, since related individuals are more likely to live close to one another than are unrelated individuals. The most obvious example is the level of sacrifice parents make for their children, who carry their genes forward into the next generation. Reciprocity is not complete here: children usually venerate their parents but in general are less willing to sacrifice on their behalf. The evolutionary explanation is that parents, even if not past childbearing, are still likely to be less productive in the future than their children will be, especially given that they already have produced their own children. Both the children and the parents have a common interest in the children's productivity, which is usually just fine with the children and naturally contributes to a belief that the world revolves around them. As they age and become parents themselves, they fall back on altruistic behaviors in the service of their genes, along with admonitions of the importance of honoring one's parents.

Parents and children are not the only ones to exchange altruism; any genetic bond can reinforce such behavior, although the degree of altruism depends on the closeness of the relationship. Brothers and sisters share as many genes with each other as they do with their parents. Altruism among siblings is strong, but a bit less so than a parent exhibits for a child. Siblings are younger than parents and would be risking more by altruistic behavior. Cousins also share genes and so tend to look out for one another, but the bond is obviously weaker than among siblings; and so on. One of the most intriguing examples of apparent altruism is provided by the Hymenoptera (bees, wasps, and ants). In these species, males have no fathers; that is, they are the results of unfertilized eggs. These children with single parents thus are ***haploid:*** they carry only the single gene copies that they receive from their mothers. Hence, when they consort with females to produce daughters, all of those daughters receive the same genetic dowries from their father, rather than the random sample that males of ***diploid*** species such as humans give to their children. The result is that sisters are very closely related, differing only in about half the genes that they inherit from their mothers; in other words, sisters differ from one another in only about one-quarter of their genes. (Actually, they differ in less than that, since in many cases both genes that the mother has to choose from are the same anyway. But that is another story.) With such highly shared genetic endowments, females of these species have much to gain from the reproductive success of their sisters. This has encouraged the evolution of very strong forms of altruism, including forgoing reproduction entirely in order to help raise the sister's offspring.

It is clear that tight links and short reward feedback loops encourage the development of altruistic behaviors; more generally, the neighborhood of interaction one has affects all sorts of ecological relationships, and hence their evolution. It is also clear that, in some cases at least, the potential or realized benefits of coalitions and other forms of altruism can encourage the formation of such assemblages, and the development of religions and systems of laws and customs that reinforce these linkages. The study of cultural evolution—of how cultures form and are propagated or extinguished—has therefore become an important complement to the study of more genetically based evolutionary theory.[20]

More generally, as systems develop, networks of interaction develop for a variety of reasons, some simply having to do with chance and geography, others having to do with choice and calculation. Plants interact with a restricted set of plants that happen to exist in their immediate neighborhood; fish and children interact with others in their school; diplomats and politicians interact with others of their ilk. Such interactions build networks, and such networks influence events. But how do those events feed back to affect those patterns of interaction? The answer, of course, depends on the circumstances. Self-organization may be a process driven largely from the bottom up, so that the patterns that we see at broader levels simply emerge from actions best understood at lower levels, or it may be a process that is strongly driven from the top down, in which individual behaviors and patterns of interaction reflect payoffs that are more diffuse and more grandly applied. For biological systems, examples of both can be found. Most of the rest of this book explores case studies in natural systems, and lessons derived from them, in an effort to gain understanding of the degree to which the patterns we see are ones we had to see, and the degree to which they are simply the results of accidents of history. The final chapter draws on these examples, and from the corresponding insights about human behavior, to explore strategies for preserving our environment.

3

SIX FUNDAMENTAL QUESTIONS

The first two chapters have posed a set of challenges, from which emerge six fundamental questions. This chapter is a summary of those questions, and a passageway to their examination. As such, it is a point of reference for the rest of the book, and a connector between its two parts. In the traditional Passover Seder, a pivotal set of questions truncates the dialogue, interrupts plans for the repast, and sets the stage for the rest of the night. The questions in this chapter are meant to serve a similar function, though the analogy is offered with humility.

The Passover questions are all subthemes on the single central question: "Why is this night different from all other nights?" Similarly, the questions to be addressed here all derive from a single question: "Why is this organism different from all other organisms?" More particularly, as ecologists and evolutionists are wont to ask, why are there so many different kinds of organisms, and why are there not more? Do ecological systems function better in some sense with more species? And if they do, would that have anything to do with how many there are, or do species simply keep evolving to exploit every conceivable niche until there are no more to be found? These are the issues that, in one form or another, energize almost all fundamental studies in ecology and evolutionary biology and from which arise the set of fundamental questions that will structure our intellectual journey.

I present here the six questions, revisit them in more detail later in the chapter, and then return to address each in turn in the subsequent chapters. Each one has a chapter all its own, except the first two, which are dealt with as a pair because they are so tightly linked. The flow of the questions follows what, for me, is a natural succession. Science begins from observation and interpretation, then seeks explanations and mechanisms. The theoretical framework that emerges provides a device for exploring the implications, and in particular for anticipating the future. Hence, it seems logical to ask:

What patterns exist in nature?
Are these patterns uniquely determined by the local environment, or has history played an important role?
How do ecosystems assemble themselves?
How does evolution shape these ecological assemblages?
What is the relationship between an ecosystem's structure and how it functions?
Does evolution increase the resiliency of an ecosystem?

I begin by examining and classifying existing patterns in nature (Chapter 4). This sort of systematic exploration provides a framework for assessing the degree to which these patterns are predetermined by physical features of the Earth and the degree to which they are self-organized. In Chapter 5, I explore the process of self-organization, and how ecological communities become assembled, by sampling the world's vast biota. Chapter 6 takes the broader and longer view, examining the evolutionary determinants of global biodiversity and the interplay between the ecological and evolutionary scales.

Ecological systems all self-organize to some degree, within the constraints that the nonbiological *(abiotic)* world imposes. That self-organization results not only in the assembly of a unique collection of species but also in an arrangement of those species into an integrated system that cycles nutrients, processes energy, and provides some degree of environmental stability for the residents. Understanding what makes a set of species a system is like understanding what makes a collection of superstars a team: one part complements another so that the whole is more than the sum of its parts. Chapter 7 asks how the dynamics of an ecosystem depend on its components, and in particular on the components' relationship to one another. In so doing, it also addresses the issue of what happens to an ecosystem if some species are lost, or if there are any changes in its internal organization. This is perhaps the central question in the current biodiversity crisis, because it points to the need for finding ways to measure and value biodiversity that reflect its importance to humanity.

Chapter 8 puts this discussion into an evolutionary context by asking whether evolutionary forces have indeed worked to increase the homeostatic properties of ecosystems, or whether systems naturally move to more and more fragile states. In particular, this can help us understand whether evolution is likely to buffer ecosystems against the new insults humans are imposing on them, or whether evolutionary trends make maturing ecosystems ever more susceptible to disturbance. Obviously, this question has deep implications for how we must manage natural systems. If ecosystems have lots of extra rivets, perhaps we can be more cavalier in our attitudes toward them, but if instead we view them as having the potential for collapse, we must take a more cautious approach.

The agenda has now been set, and it is a full one. Before beginning the search for answers to these challenges, however, I will endeavor to justify in a bit more detail why these questions capture the essential issues.

Question 1: What Patterns Exist in Nature?

The facts are always a useful place to begin. The American humorist Kin Hubbard said, "Nobuddy kin talk as interestin' as th'feller that's not hampered by facts or information." It is certainly my predilection to focus on abstraction and inductive reasoning. Obviously, some balance is in order, however; it is essential to ground theories to the greatest extent possible in facts, and then to evaluate the facts within the framework that theory provides. Thus, before speculating on what patterns we *should* see in nature, it seems a good idea to ask what patterns we *do see* in nature. More specifically, the question is, What patterns exist in the distribution and organization of biodiversity?

Science is, first and foremost, a way of knowing about the natural world. Hence, it takes its roots in observations about nature rather than in axiom systems. But science also requires generalization and seeks to interpret countless infinities of unique phenomena in terms of the smallest possible number of fundamental forces and explanations. It is not enough to report that an apple fell from a tree and that a book fell from a shelf; science seeks to recognize these facts as different manifestations of the same basic process, arising primarily from the same force: gravity. That such insights can also help to explain why the moon stays in orbit around the Earth, or the Earth around the sun, profoundly increases the power of this scientific insight to explain the world around us.

Thus, the key to scientific inquiry is to take apparently disparate events and find ways of arranging and explaining them within parsimonious structures of data and theory. The delineation of broad patterns in the distribution of species became the focus of study by ecologists and evolutionists even before such professions had those names. The simple description of a wide variety of systems, while useful, would not by itself advance the science a great deal. However, through careful analyses, scientists have managed to arrange data into categories, aggregating agents into meta-agents and then relating those entities to one another and to physical factors in order to provide a systematic framework for organizing information.

The periodic table of the elements has made it possible for chemists to develop a comprehensive science, grounded in physical principles; the Linnaean classification of species did the same for the study of nature. In a similar way, the systematic organization of the distribution of organisms with respect to physical variables such as soil and climate is the essential first step in the development of a science of ecological organization. It is easy to find characteristic patterns of temperature and rainfall for any region of the globe in reference books. Along with a decent atlas, one can then make pretty good guesses about where, for example, to find tundra and where to find rain forests. This provides a first cut at classification, allowing the subdivision of the globe into *life zones.* It is only a first cut, though. Rain forests will not occur where they can't, but they will not always occur where they can, perhaps because they have never been given the opportunity to try. That is, limitations to dispersal may prevent potential colonists from ever reaching habitats where they could do very well indeed.

Ecological studies, over the last half-century especially, have deepened our understanding of the relationships of species to environments: where deserts will occur, what the most productive regions of the oceans are, and so forth. Factors such as temperature and humidity delineate the types of plants, the types of animals, and hence the types of communities and ecosystems that can exist. Similarly, physical factors determine the patterns of productivity in the oceans, and thus the distributions of marine communities. Within those limitations, there is considerable scope for variability, but a careful examination of what is known provides a framework for asking more speculative questions.

Chapter 4 explores what is known about how plants and animals are distributed over the globe, and why things are where they are. This is more than a travel guide: it leads naturally to asking why, say, marsupials are primarily an Australian specialty and not more widespread. Or why Australia and New Zealand are not more similar in their biotic composition. Or why San Salvador and Bombay, which have quite similar ranges of temperature and humidity and similar oceanic exposure, have very different collections of plants and animals. One answer to all of these questions is that these areas have had unique histories and are surrounded by other areas that have had their own unique histories. Space, chance, and history hold the key. And that is why Chapter 4 must also address . . .

Question 2: Are These Patterns Uniquely Determined by Local Environment, or Has History Played a Role?

More precisely, this question should be: To what degree are patterns determined by local physical conditions, such as soil and climate on land, or salinity and flow patterns in the ocean, and to what degree are they self-organized? It is here that nonlinearity comes into play. Natural systems are highly nonlinear, and what we observe in any environment is in part the result of accidents of history and of the influence of emigrants from neighboring ecosystems. Evolution is a tinkerer, in Jacob's analogy; moreover, what we currently see in nature, at any level of detail, is (as Stephen Jay Gould has elegantly discussed) not what we would see were the tape run a second time; that is, if the evolutionary process were restarted from its primitive conditions.[1] Nature's cast does not consist of optimally adapted organisms, uniquely determined by local environmental conditions.

This simple but fundamental point frustrates efforts to predict how forests or grasslands, for example, will respond to global climatic change; we can't just read the answers off a map or a chart because space, chance, and history matter too much. As we shall see, broad features of vegetation may be predictable, but the exact composition is not, even given every possible detail about temperature, moisture, topography, and soil conditions. It matters little whether our perspective is an ecological one or an evolutionary one; both involve differences of degree rather than of kind.

Explanations for how patterns are generated and maintained vary depending on the scale imposed by the commentator, and a full picture requires integrating and matching explanations across scales. Quite typically, factors extrinsic to the system of study hold sway at certain scales, while intrinsic factors govern patterns at others. That is, patterns at certain scales are largely imposed, while those at other scales represent self-organization. Animal schools and herds provide wonderful examples. Environmental conditions determine where animals of a particular type can occur at all and thereby impose broad distributional patterns; within a range, however, animals control their own grouping patterns through their responses to one another, as well as to environmental cues such as the distribution of food.

In terrestrial environments, such as the Serengeti in Kenya or the vast plains in Northern Europe or North America, the interplay between physics and biology operating on different scales produces a wide range of grouping patterns in species such as wildebeest, reindeer, and bison. Each can be found in herds of various sizes and geometries, the details reflecting how self-organization has operated as a result of particular local conditions and particular accidents of history. There are evolutionary reasons for animals to be in groups—grouping can aid in the finding of resources, in escaping predation, or simply in finding mates—but the specific grouping patterns are not so important. The great herds of Serengeti wildebeest, for example, which arise from the collective movements of many thousands of animals each paying attention only to its local environment and to what its neighbors are doing, thus exhibit many different shapes and forms.

FIGURE 3.1. Aerial photograph of a large wildebeest herd. Courtesy A.R.E. Sinclair and University of Chicago Press (plate 3 from Sinclair, *The African Buffalo* [Chicago: University of Chicago Press, 1977]).

In the oceans, patterns of circulation and currents cause patches of water to become isolated and coherent rather than mixing freely with the rest of the ocean. These give rise to aggregations of fish and *invertebrates*, as well as of the tiny plants (phytoplankton) that lie at the base of the food chain. Within those patches of animals, smaller patches form for entirely different reasons: individuals, to some extent, need each other's company, and their mutual attractiveness organizes them into small patches within the large patches.

It requires no stretch of the imagination to realize that human groupings are governed by similar sorts of extrinsic and intrinsic rules, at different scales. Basically, we must live on dry land, most preferably (but not necessarily) near large water bodies that allow the easy flow of goods and services. These basic principles impose overriding distributional patterns; but within these, smaller groups form, owing to human behaviors. Individuals like to live near other individuals in order to reap the benefits of community life and the division of labor. Some individuals prefer smaller communities, some larger ones; but even small communities tend to grow and to develop connections with other small communities, creating larger meta-communities. Spatial structure at a variety of scales thus develops.

Chapter 4 explores in greater depth the interplay between necessity and self-organization in the structure of ecosystems. As this brief discussion has shown, the details of that interaction are not uniquely specified but vary with the scale of investigation.

Question 3: How Do Ecosystems Assemble Themselves?

Ecological and evolutionary processes are similar in kind, distinguished primarily by the distinct time and space scales over which they operate. This difference in scales, however, makes it essential to separate ecological and evolutionary events from one another and to study the interplay between them. Ecological interactions take place within an evolutionary context and in turn shape the ongoing evolutionary process. Interactions between species lead to the local replacement of one by another, or at least to differential success that influences the relative abundance of types. Communities communicate with one another through the exchange of individuals. The collective dynamics of communities lead to biogeographic patterns in the distribution of species and to the determination of the global inventory of plants and animals, the Earth's biota. In turn, the assembly of any community following a disturbance, whether a volcanic eruption or the death of even a single tree in a forest, involves the biased selection of species from this global biota.

Chapter 5 examines the assembly rules by which communities are formed from the effectively infinite range of possibilities. If there are N species in the world, one could in principle form communities in which any of those N species would be present or absent. With N such binary decisions to be made, there are 2^N possible branches to the decision tree. One branch results in a community with no species, but that still leaves 2^N-1 possible communities that could be constructed from those N species. This simple calculation deals only with the presence or absence of each species and hence does not take into consideration variation in abundances. Thus, it obviously grossly understates the number of communities that could be formed but provides us nonetheless with a convenient place to begin.

By this calculation, if there were only ten species in the world, there would be more than one thousand possible communities that could be formed from them. With as few as twenty species, there are over one million possibilities. With one hundred species, the number of possible communities is astronomical. Since there are really not ten, twenty, or even one hundred species in the world, but on the order of ten million, the number of possible communities is far beyond what we can fathom. So what determines which communities do arise, and which don't, from this unmanageable diversity of possibilities?

First of all, environmental variation limits where species can be and restricts the range of possible types that can persist in any environment. But the number of remaining possibilities clearly still is off the scale, since we know that the thousands of species to be found in any environment are only a small subset of those that could survive there if introduced under favorable conditions. Evidence for that comes from the sorry history of the introduction of exotic species, countless numbers of which have become pests in their new habitats: gypsy moths, Mediterranean fruit flies, killer bees, kudzu, purple loosestrife, walking catfish, rabbits, zebra mussels—the list of introduced species that have flourished when transported to a new habitat is endless. Hawaii, in particular, is a virtual mishmash of introduced species, which dominate the island landscapes. So environment helps explain the patterns, but it is only a part of the story.

History also is vitally important. Over the course of the evolutionary development of the world, there have been uncountable junctures at which any of a large number of species could have made their way into any of a large number of communities. Accidents of history have determined which have made their appearances, to some extent limiting opportunities for other species (and to some extent creating new opportunities for yet other species).

Philosophers and developmental psychologists tell us that human development is the closing of doors of opportunity. A newborn baby is a blank slate, with infinite possibilities for intellectual growth. Events in early childhood development take the individual down certain paths, restricting the potential for future choice. The same is true of our developing planet, which is home to the particular collection of species that share it with us in part because of physical constraints, and in part because of accidents of history that have become frozen and reinforced. This is nonlinearity at work. This is self-organization.

History has restricted the composition of any ecological community, as well as the composition of its neighbors. When opportunity for colonization presents itself within an ecosystem, a limited set of colonists apply for admission; it is a case of being in the right place at the right time. And from those applicants, only those that can integrate themselves into the existing community will succeed. Pest species find such opportunities, usually with the help of humans, when they get to places where they have not been before. More generally, the new colonists are species not so different from those that have been there before, because neighboring communities, which supply the potential colonists, are likely to have similar biotas.

When broad-scale disturbance occurs, as, for example, following a volcanic lava flow, colonists are drawn from a larger geographic area. Still, there are strict rules governing the successional recolonization of the newly exposed surfaces. The first colonists must be *autotrophic* (photosynthetic, or otherwise able to manufacture their own food from inorganic materials), that is, capable of fending for themselves and deriving sustenance from a harsh environment. These in turn prepare the way for others, in a reasonably regular pattern of successional development. Within the constraints set by environment, history, and the composition of neighboring environments, ecological communities self-organize according to a roughly repeatable set of rules. The interplay of all of these factors provides us with insights into why certain communities exist; the challenge is to determine the relative importance of these various influences.

Question 4: How Does Evolution Shape These Ecological Assemblages?

The subtle interplay between ecological and evolutionary events represents the integration of processes along a continuum of scales rather than a dialogue between two sharply distinguished ones. We tend to think of ecological dynamics as occurring over time periods of days to years, and of evolutionary events as occurring over millennia. This assumption, however, is a gross and misleading simplification. Important evolutionary change, such as the development of pesticide resistance in weeds or of antibiotic resistance in bacteria, can occur in a very small number of generations. The very large differences in life spans among species that tightly interact within ecosystems means that the evolutionary time scale for some species, such as bacteria and viruses, may be as short as or even much shorter than the ecological time scale for other species. Thus, bacterial resistance to antibiotics has become a frightening and major public health problem in our lifetimes; plant breeders are in a continuing battle to develop new crop strains that can fend off rapidly evolving fungal pests; and efforts at pest control have been continually frustrated either by the development of resistance to chemicals or by the loss of effectiveness of biological control agents. The classic example of the last challenge, the limited success of the *Myxoma* virus in controlling rabbit pests in Australia and elsewhere, is a case to which I return in great detail later.

The coincidence of some ecological and evolutionary time scales is not just a consequence of the interactions among species with very different life spans. Individuals are organized into groups, and those groups into larger groups. Individuals grow, reproduce, and die; so do groups. The time scales over which these events occur are very different and thus create evolutionary changes over a range of scales. Many groups are loosely defined, transient entities with little influence on the evolutionary process; sometimes, however, the influences are strong, and group-level processes can exert important pressure on the evolution of individuals. Indeed, we will learn that organisms themselves are not the fundamental elements of the process but are simply coalitions of genes that have their own, more rapid dynamic.

This exploration of scales demonstrates that the dynamics of any ecosystem involve a continuing panoply of ecological and evolutionary interactions among a diversity of agents and meta-agents. The ecosystem is a complex adaptive system whose development over a period of years involves selection among its components over much shorter periods of time, while also reflecting other evolutionary processes that have occurred over much broader space and longer time scales. A forest is a complex of species, some adapted to the slow growth and eventual dominance of the canopy, others adapted to rapid exploitation of the temporary gaps that form. When a tree dies in the forest, or when a cluster of trees is felled by windthrow, a local successional process begins. The early colonists are adapted for rapid dispersal and growth; these include species such as grasses, forbs, and shrubs that require and thrive on lots of sunlight. As the forest develops, light availability in the understory diminishes, and these species are replaced in succession by those that make their living not by opportunism but by their ability to grow under low-light conditions and that hence will prosper in a highly competitive environment. Is the forest landscape so different in this way from the economic landscape, in which opportunistic entrepreneurs leap in to exploit newly available possibilities, ultimately to be replaced by other companies better suited for the long haul?

Ecologically, specific traits adapt a species for exploiting a particular ecological niche; evolutionarily, the availability of new ecological opportunities creates pressure favoring their exploitation. Evolution hates a vacuum. That is not to say that every niche will be filled, or that a list of niches arrives in the same box with a newly minted ecosystem, with instructions on how to assemble ("Place hex species cc in serrated niche x7, being sure to keep edges reversed and ecosystem flat"). As a system develops, new opportunities are created as parasites or predators find new hosts and prey to exploit, or as pollinators and hosts develop their mutual interests. In this way, ongoing evolution changes the adaptive landscape for other species; this is a particular manifestation of the system's nonlinearity, and it reinforces the importance of historical accidents that irrevocably influence the system's later development.

Over the evolutionary time scale, a similar dynamic applies to ecosystem types; a hardwood forest in the northeastern United States, for example, is just one unit of a meta-forest, a collection of forests with similar but not identical compositions. Following a major disturbance, the reassembly of a forest is not a novel occurrence; it takes place within the context of that meta-forest, which provides the evolutionary background. The evolution of the northeastern forests as a group, in turn, represents the collective dynamics of a multiplicity of individual forests.

The interplay between ecological and evolutionary processes is central to understanding the emergence of biodiversity, a topic to which I will return in discussing the next question. Forest evolution involves processes identical in kind to those that determine the characteristics of the species that make a living on lava flows, or the intertidal invertebrates that exploit wave-induced gaps in mussel beds. Chapter 6 explores this issue in detail and is in many ways the central chapter of the book.

Question 5: What Is the Relationship Between an Ecosystem's Structure and How It Functions?

I have already emphasized that the loss of biodiversity threatens the maintenance of a wide range of goods and services that humans derive from ecosystems. Such benefits range from the food, fuel, and fiber that we harvest to the maintenance of climate and clean air and water. Not all species or components of biodiversity are equally important in maintaining those features, leading some to suggest that *conservation biology* must focus on developing a version of *triage:* giving the greatest attention to the most vital components. But to be able to develop such a scheme requires an understanding of the roles that individual species, for example, play in maintaining system properties. How should we go about acquiring that understanding? This is the topic of Chapter 7.

An electronics expert can look at the wiring diagram of a stereo or a television and understand how the receiver works; an automobile mechanic similarly can deduce how the automobile operates by examining its parts and how they are hooked together. Such diagnostic abilities, of course, are possible because these devices are built according to standard plans, with only minor variations; furthermore, the principles of design have been shaped exactly to achieve specific functions. To a large extent, these principles extend to the design of organisms; for example, a doctor's principal tools of analysis involve examination of the parts, directly or indirectly.

Ecosystems are different. Their dynamics emerge *from* the pieces and their interactions, rather than guiding the principles of design and assembly. Indeed, ecologists, eager to make this distinction clear, generally prefer to talk about the functioning of an ecosystem rather than the function or functions; ecosystems, from an evolutionary perspective, have no function, or at least they have not evolved to fulfill specific functions. This does not mean that they do not serve functions; they do. The species that exist within ecosystems rely on the environment that those ecosystems provide and, of course, to some extent help define that environment; but this is very different from suggesting that the ecosystem has been designed to have certain properties in order to make it possible for particular species to exist. Fishermen live on coasts because that is where the fish are; the fish do not come to the coast in order to accommodate fishermen.

This fact complicates any analysis of the relationship between the structure of an ecosystem and its functioning, including the services it provides. In the absence of design for purpose, one lacks the handbooks that explain what individual parts have to do with overall system dynamics. But ecosystems have important properties that *emerge* from the adaptive evolution of their components; emergent processes include the processing of energy and materials and the mediation of climate, all of which are important to humans as well as to all manner of plant and animal life. It is in principle not difficult to measure the flow of energy through the parts of an ecosystem and to add the flows together to give a static picture of ecosystem processes. But such a description gives little insight into system functioning, just as tracing the flow of food as it is processed in the body gives little insight into what makes an individual tick. The health of an individual is governed by homeostatic mechanisms, feedback loops that keep critical processes on target. Overexertion brings exhaustion, causing an individual to slow down; thermostats in the brain stimulate the secretion of hormones to regulate the production of heat in body tissues. Introduction of an alien agent, an antigen, into a vertebrate activates an immune response, which produces antibodies to eliminate or control the antigen. All of these are adaptations, feedback responses that help maintain the internal environment.

In a similar manner, there are homeostatic mechanisms that regulate ecosystem processes, maintaining their resiliency in the face of perturbations. Unlike the homeostatic mechanisms in the body, homeostatic mechanisms that maintain ecosystem processes have not been shaped by selection in order to maintain those processes; rather, the processes that emerge are those that exist because there happen to be homeostatic processes to maintain them. In a sense, it is the processes that have been selected rather than the homeostatic mechanisms. Nonetheless, understanding how ecosystems work requires understanding those homeostatic mechanisms, and doing so requires, in turn, studying systems far from *equilibrium*.

Far-from-equilibrium dynamics impose an important requirement on the elucidation of ecosystem behavior, and of the relationship between structure and functioning: the experimental manipulation of natural systems. Eugene Likens, Herbert Bormann, and their colleagues, for example, by establishing various cutting regimes in forests in the Hubbard Brook Watershed, have learned about the effect of practices such as clear-cutting on a forest's ability to maintain its functioning.[2] From studies such as these, and from the work of other investigators such as Robert Paine in marine intertidal regions, David Schindler and Stephen Carpenter in lakes, and David Tilman in experimental plant communities, a picture has begun to emerge of how ecosystem functioning depends on the details of biotic structure.[3] Such studies allow us to begin to evaluate how to measure biodiversity in terms of its importance to maintaining ecosystem goods and services.

Question 6: Does Evolution Increase the Resiliency of an Ecosystem?

The fourth question asked how evolution shapes community properties, and the fifth, how the structural and functional features of ecosystems relate to one another, and in particular how ecosystem structure affects the resiliency of its processes in the face of perturbations. One specific aspect of this challenge is to understand how the diversity within a functional group buffers the functioning of that group against stresses. Just as a genetically diverse species is better able to respond to environmental challenges, so, too, may a genetically diverse functional group be expected to be more capable of responding to challenges at a higher level.

Many conventional theories suggest that there is selection for generating and maintaining genetic variability within populations, and in particular that sex and the associated genetic recombination are maintained in fluctuating environments because they enable the genome to maintain the variability needed to respond to changing conditions, including temperature variation as well as the spread of parasites and infectious diseases. Natural selection is a process that destroys variation, preserving the most fit from among pretenders; environmental change, however, puts a premium on the maintenance of the variation necessary for continued evolution. At the organismic level, this evolutionary tendency can lead to second-order selection for reassortment and novelty, in the form of recombination and mutation. The tension between the forces that generate variation and those that dissipate it results in a compromise in which intermediate levels of variation are maintained. But to what extent are such forces at work at higher levels of organization, such as the species, the functional group, or the ecosystem? A functional group is much less tightly organized genetically than a species, and a species is less organized than a collection of genetically related individuals. What happens to variation within these groups over evolutionary time, as well as over the time scale of the development of an ecosystem? Competition between species can lead to *competitive exclusion,* and hence to the loss of diversity. Are the compensatory forces operating at levels above the species sufficient to maintain diversity? Or are there other processes, such as chance and history, that endogeneously generate and sustain diversity, without major selective influences flowing down from the system level?

Per Bak and his colleagues have suggested that self-organizing systems drive themselves to *criticality,* producing accidents waiting to happen, sandpiles at the perpetual edge of collapse.[4] The term "criticality" is borrowed from physics: a critical temperature is one that characterizes a *phase transition,* such as that between solid and liquid. Critical temperatures for water, for example, are the freezing point and the boiling point. If ecological systems evolve to critical points, then it would suggest that functional groups have no buffering capacity, and that ecosystems are forever on the verge of catastrophe. Most ecologists would dispute this, arguing that there is some redundancy within functional groups, such that shifts within groups would have little effect on system processes. The idea is to some extent testable, because the hypothesis that systems are at the edge of criticality implies that we should see cascades of catastrophes obeying characteristic probability distributions, in which the frequency of a particular level of disturbance decreases as a constant power of the level of severity. Earthquakes, for example, seem to fit such distributions; in particular, there are many more small earthquakes than large ones, and the pattern of decay in frequency fits well a distribution of the *power law* type.[5] Do such laws apply to catastrophes hitting ecological systems, or extinctions striking taxonomic groups? Despite efforts to test this idea, for example, for extinctions in the Phanerozoic era, the hypothesis remains unsubstantiated. However, it certainly represents one of the most intriguing and tantalizing open research problems, with tremendous implications for our understanding of the nature of ecological systems and the influence of evolutionary processes.

Stuart Kauffman, among others, argues that Per Bak's self-organized criticality is manifest in an evolutionary tendency for ecosystems to reach the *edge of chaos*—a world in which extinction and *speciation* are constantly occurring, maintaining the variation needed for adaptability—and that it is in this precarious balancing that ecosystems achieve a form of stability.[6] This theory, in its rawest form, would be anathema to an evolutionary biologist, because it suggests that selective processes operating at the level of ecosystems maintain the same sort of variation that mutation and recombination sustain at lower levels of organization. Without identification of a mechanism for keeping an ecosystem at this edge, it is a difficult hypothesis to swallow. Kauffman understands this and has developed and encouraged exploration of mechanistic models that might support the contention that evolution to the edge of chaos is a universal property of self-organizing systems.[7] Indeed, Per Bak and his colleagues have demonstrated such tendencies in some highly idealized models, but ones that suppress much of the similarly self-organized complexity of ecosystems.[8] I return to explore this further in Chapter 8.

Thus, the challenge is set. How has biodiversity arisen? What maintains it? And how fragile is it and the services it provides? The chapters that follow explore the answers to these fundamental questions, providing a bridge to understanding how to protect ourselves from the loss of those services.

4

PATTERNS IN NATURE

The fundamental challenge in understanding the organization of any complex system is to sort out the role of history. Which of the features that we see could have been no other way, given the influence of the local environment? Which, on the other hand, were determined by chance, accidents of history at critical junctures of a developmental process that offered multiple alternative pathways for exploration? If our focus were child development, the issue would be the familiar one of nature versus nurture, of genetics versus environment. In the case of ecosystem patterns, it is an issue of necessity versus accident, as reflected in the first two questions from the master list. Hence, before we begin to explore possible *endogenous* mechanisms for patterns—that is, those that represent the self-organization of ecosystems—it is important that we understand just how much scope for variation there is in the process. Natural systems, after all, like any systems, must obey the laws of physics and chemistry. These restrict the development of ecosystems by limiting where particular biological forms can exist as well as by imposing thermodynamic constraints on energy exchanges. We will discover in this chapter that such constraints determine where rain forests can occur, where deserts will be, and where grasslands are to be found; but that, within those broad constraints, there is a great deal of latitude concerning the exact species that may be found.

The Possible and the Actual

In the biblical creation story, the fish are put in the sea, the fowl in the air, the cattle and creeping things upon the Earth. Good ideas all—imagine what would have transpired had the fish been put on land, the cattle in the air, the fowl in the sea. Not much would have survived except for a few seabirds, dodging falling cows. Each organism has its place, shaped by evolution to meet the constraints of the environments to which it has adapted. This principle allows ecologists to classify individuals according to habitats, and habitats according to the types of species that can live there. Such classification provides only the broad outlines of explanation. Within any habitat, only a small proportion of the species that could exist there are found, since colonization accidents and subsequent interactions between species have shaped the actual assemblage. In the same way, and for the same reasons, most species are found in only a small fraction of the habitats where they could thrive.

The late Yale University ecologist and limnologist G. Evelyn Hutchinson, the most important figure in American ecology, illustrated this dichotomy by distinguishing a species' *fundamental niche*, the range of conditions under which it could survive, from its smaller *realized niche*, the range of conditions in which it is actually found. The term *niche* confounds two concepts, the range of physical conditions in which a species may be discovered and the spectrum of ecological roles it may play within a community. For the present discussion, it is the first aspect that is of primary interest because the goal is to understand what the external constraints on evolutionary processes are.

The dichotomy between the fundamental and the realized mirrors the duality that François Jacob called *the possible and the actual*. Indeed, the title of this section is borrowed shamelessly from the title of his 1982 book, which I heartily recommend for its perspective on evolution.[1] Hutchinson was concerned primarily with processes in ecological time, using the terms "fundamental" and "realized" to distinguish a species' potential niche from the roles and places it manages to fill. Jacob's attention, in contrast, was on the evolutionary time frame. "Possible," then, referred not just to extant species but also to hypothetical ones that are absent from the real world—hence, not among the "actual." The differences between Hutchinson's concerns and Jacob's are subtle, but worth preserving through the separate terms. The similarities between the concepts, however, are basic to science and deserve deeper exploration.

The distinction between the possible and the actual, between the fundamental and the realized, highlights a dilemma we face in trying to make any predictions about future evolutionary change—for example, in endeavoring to anticipate what projected climate change will mean for forest communities and other aspects of the world's biodiversity. The vegetation that exists under any particular set of conditions reflects only the realized niches of the species that are there. That is, in an extension of Hutchinson's profound concept, the species to be found in any community represent its *realized biota* rather than its *fundamental biota,* the complete set of species that could survive under the range of soil, climatic, and other physical conditions that prevail there. Since it is impossible to infer what the hypothetical fundamental biota is simply from information about the much more limited realized biota, it is consequently impossible to predict what the realized biota in any new environment will be. Indeed, even if we somehow knew the fundamental biota exactly, perhaps from experimental studies that expose physiological limits, it would still be impossible to determine exactly what subset would form the new realized biota. The specifics of what species would be found would remain unresolved: Would species that could potentially survive ever reach the new habitat? And even if they did, how would their success in becoming established be affected by the other species that got there first? These are questions about the dynamics of colonization; that is, they go well beyond what can be learned from statics. One can gain some insight into resolving them by constructing and exploring mathematical and computer models that are predicated on knowledge of fundamental mechanisms and interactions. Such models, however, are not without their own problems; hence, they can reduce our ignorance somewhat but never eliminate it entirely.

Circumstantial Evidence

Unraveling the details of why species are found where they are has been one of the most exciting detective stories in ecology. It is, perhaps, the central problem. Without dynamics, we must turn to correlations—circumstantial evidence. Inference from such information certainly has its weaknesses but provides a potent place to begin formulating hypotheses. The fact that the owner of a building took out a fire insurance policy on his building the day before it burned down does not prove malfeasance, but it makes for an interesting coincidence. It even is likely to suggest some intriguing hypotheses to the insurance company and the local gendarmes seeking to solve the mystery of why the fire began. Correlations establish connections and help guide the search for explanations.

Ray Hilborn and Marc Mangel, two very clever mathematical ecologists, recently published *The Ecological Detective,* a book that, as its evocative title suggests, tells the story of how ecologists can use models and statistical methods to explore hypotheses and solve ecological whodunits.[2] Like other detectives, ecologists find it useful to begin from correlations, especially those between organisms and environment, but also ones involving the coincidences of species within particular habitats.

Correlational approaches can be quite problematical, however. Coincidence of species within the same habitat may mean that those species share similar resources, or it may be that they exploit different resources and coexist because they have found ways to live side by side without engaging in excessive competition. The plot thickens. For example, environments may be characterized broadly by mean temperature and moisture, but these gross measures smooth over the smaller-scale variations that species have learned to distinguish. Thus, coarse descriptions of habitats at only the level of macroscopic or average statistics can suppress much of the important fine-scale detail, in space as well as in time, that is essential to understanding patterns of species coexistence.

A second problem is evident in efforts to use correlations to infer ecological compatibility or incompatibility of species. Since any local biota represents only a sample from a much larger fundamental biota, the fact that two species do not exist in the same habitat tells us nothing definitive about whether they could. Much of our understanding of community assembly processes has come from the study of islands, and especially groups of islands like the Galápagos that share at least some common physical and biological features. When nearby islands are found to have different biotas, explanation can be sought in terms of physical differences between the islands, such as their size, their ruggedness, or their topographic complexity; in terms of their degree of isolation from mainlands or simply the vagaries of colonization; or in terms of the compatibility of new arrivals with species already present. Given the fact that, with a mainland pool of even as few as 100 species, there are more than 1,000,000,000,000,000,000,000,000,000,000 (one nonillion) possible communities that could be assembled, any process of assembly obviously produces only a tiny subset of the range of possibilities. This does not mean that deterministic rules of assembly have played no role in shaping the realized biotas; of course they have. But it does caution against reading too much into the absence of a species from a particular location; there is no way that we could observe more than a small sample of the spectrum of possibilities.

Nonetheless, there is much to be learned from correlative studies concerning the fundamental biotas of particular habitat types; it is a way to understand the scale at which reliable predictions can be made. Alpine regions in which both annual precipitation and temperatures are low is home to tundra and to those species that are characteristic of tundra. Hotter, drier areas host deserts and desert species; hotter, wetter areas are occupied by rain forests and rain forest species.

Taking particular climatic conditions as given, and assuming they drive ecological assembly, has an inherent problem: climate is not independent of community composition. That is, as communities change in response to changing climate, there is a feedback between vegetation and climatic conditions. Desert conditions can arise, for example, if overgrazing removes the vegetation that would otherwise return moisture to the atmosphere. That change affects predictions of future climatic conditions, but it does not alter the fact that desert species are found in deserts, and that deserts are home to desert species.

The defining characteristic of the fundamental plant biota of the desert is an ability to survive under conditions of high temperature and high water stress. The cactus, a specialist for desert conditions, has replaced leaves with stems that reduce water loss and thus has developed the capacity to store water. Many annuals exist in the desert, primarily therophytes; in these, seeds are more easily protected from extreme temperatures than the above-ground parts of adult plants and thus provide a sensible way to wait out harsh conditions. Perennial species in the desert are characterized by high investment in root systems that help draw water from the ground and expose as little as possible of the plant to harsh above-ground conditions. Many plant species found in places like Death Valley, California, are hemicryptophytes, which die back to ground level during harsh conditions, thereby protecting under the soil the buds that will regenerate and form the next year's growth. Hemicryptophytes, however, are more typically associated with temperate and arctic regions, where winter presents challenges for plants that are not annuals.

The classification of species into therophytes and hemicryptophytes—along with phanerophytes, chamaephytes, and cryptophytes—goes back to the efforts of the Danish botanist Christen Raunkiaer in 1903. Raunkiaer reasoned that by distinguishing plants according to where they placed their buds, one could infer the types of environments in which they could survive. The validity of Raunkiaer's logic is demonstrated by the fact that his scheme has survived in practice for a century.

By focusing on adaptations for desert conditions, we can characterize the fundamental biota of the desert. In this way, we have a good idea of what desert plant species look like. Furthermore, based on what we know about desert plant species, and what we know separately about animal adaptations for tolerance to arid conditions, we can infer a great deal about the types of animal species to be found in deserts; for example, they obviously must be able to exploit for food the unique desert plant forms. Thus, seed-eating rodents are common in deserts, because they possess the ability to make use of the available resources and because through nocturnal habits and hibernation they have evolved to avoid the most extreme climatic conditions.

Considerations of this sort have led to a variety of schemes for classifying the vegetation of particular kinds of habitats. While it is not feasible to specify the exact species assemblage in any ecosystem simply on the basis of physical variables, it is possible to say a great deal about the functional groups that one is likely to find. And the broader the spatial perspective—that is, the *scale*—the more reliable and the more detailed the predictions can be.

Scales of Perception

The issue of scale is a fundamental one in efforts to relate community characteristics to environmental conditions. General statements are easily made about the vegetation over large zones, but the fine-scale details are less easily specified. The next section discusses the broadest scale, the so-called life zones, such as tundra and rain forests, at which level quite robust predictions can be made. Within life zones, at what may be called the *mesoscale,* or middle scales, patterns exist but exhibit more variation; these are the topic of the subsequent section. Finally, at the finest scales, exact prediction is virtually impossible because of the role of chance.

These observations are not unique to ecological systems. In general, in describing the dynamics of any system, the coarser the field of vision, the easier it is to make definitive statements. Financial advisers warn that no market predictions can be guaranteed over the short term, though these same advisers confidently tell investors who are in for the long haul that equities are the way to go. Over any period of ten years or more, it has almost always been the case that stocks provide better rewards than do more conservative investments. Everyone tells me that now, and I believe them. Nobody told me that when I was younger; I assume that they knew it then, too, but maybe believed it less because the Great Depression was still a vivid memory. The same is true for weather prediction: a longer perspective allows more robust prediction.

These examples deal with time scales, but spatial scales provide perhaps even more obvious illustrations. Any large city predictably has a plethora of fast-food establishments, rest rooms, and gas stations, but try to find one when you need it in an unfamiliar neighborhood. What is predictable at the level of the city is unpredictable at the more local level. In the same way, it is more reliable to make predictions about the collective behavior of whole forests than it is about individual trees or small parcels of land, in terms of how the vegetation will develop over, say, the next fifty to one hundred years. There is in this a general principle, the foundation, in fact, of statistical mechanical theories that allow us to deal predictably with the macroscopic consequences of unpredictable local phenomena. In the simplest cases, this is just the *Law of Large Numbers,* the mathematician's way of making sense of large ensembles of statistically independent events. In other cases, the macroscopic regularities may have a deeper explanation, reflecting, for example, homeostatic mechanisms operating at some higher level of organization.

Whatever the mechanism, the point is that one's perception of any system and its fluctuations depends on the perspective assumed, the scale of observation. For many phenomena of interest, a coarser scale allows one to average over variations and make more definitive statements, whether about space, time, or the ensemble of agents in a complex adaptive system. It is far easier to make reliable predictions about demand for automobiles over the next decade than it is to make predictions about demand for Jeep Cherokees. In the same way, it is easier to make reliable predictions about total fish catches than about the catches of any particular species, or to make predictions about forest biomass than about any tree species. The chestnut largely disappeared from American forests unpredictably, owing to the sudden emergence of blight, but other species in its functional group essentially filled the gaps, leaving little obvious change in forests except for the absence of chestnuts.

A coarser scale does not necessarily guarantee enhanced predictability. I have a much better idea of what I will be doing over the next week than over the next ten years. The fact that I do not even know whether I will be alive ten years from now puts an unfortunate damper on my predictive capabilities. Averaging may not help predictability if it is taken over scales where there is a strong trend. The main message, then, is that scale matters in prediction; things come into focus, stroboscopically, on certain scales, and not on others. Moreover, the explanations for the patterns seen on different scales differ.

Recognition of this principle has guided our understanding of natural systems and led us to explore the interplay of mechanisms that operate on different scales. The variations in how I feel from day to day are driven by essentially random events. The chance that I will come down with a cold virus this winter depends largely on what is circulating in the community, though, of course, my reaction to exposure does depend on my own susceptibility. On the other hand, conditions such as air and water quality may certainly be contributing factors, and obviously the gradual decline in my physical fitness over the decades has more to do with my own aging process than with the external environment. The point is that, on short time scales, the factors accounting for variability in how I feel are very different from the ones on longer time scales; my doctor and I must synthesize these in developing an integrated view of my health. Though I am not reassured when my physician tells me that my internal systems are as good as can be expected for someone my age, I understand.

The same principle applies to the stock market. Economists and market analysts depend on the fact that short-term market fluctuations are largely unpredictable, though "Adam Smith" argues, "I suspect that even if the random walkers announced a perfect mathematic proof of randomness, I would go on believing that in the long run future earnings influence present value."[3] Short-term market fluctuations are obviously heavily influenced by essentially random events external to the market, such as congressional actions, war in Somalia, or the outcome of an election. Over the longer term, these influences tend to even out or self-correct, and market variation must be explained in other ways. If I understood these, I probably would be spending my time doing something other than writing this book, even though it is a labor of love.

Averaging over time and averaging over space are not so different in effect; thus, for example, it is easier to predict the behavior of the whole forest than the behavior of the trees. Similarly, predictability generally goes down as one divides things more and more into categories—that is, as one reduces the degree of aggregation into meta-agents. This is, after all, the central tenet that makes the insurance industry a good deal for buyer and seller alike. When I buy insurance, I am protecting myself against uncertainty associated with the fact that I and my family constitute a small sample; the insurance company, on the other hand, cares little about individual cases because it takes a broad sample. It averages over uncertainty, in the way I would by buying mutual funds, or the way a plant would by dispersing its seeds over diverse microhabitats. A similar dichotomy explains Las Vegas, though in that case many people crave uncertainty and the slim likelihood of sudden wealth.

Prediction of the dynamics of ecological systems—or of socioeconomic systems, for that matter—similarly varies a great deal depending on how one chooses to lump things together. The demand for university professors twenty years from now, for example, will predictably be driven by human birth rates now. Less predictable is how demographic change will affect the demand for professors of music, and even less so for Berlioz specialists. Demographic trends are unlikely to have much to do with interest in Berlioz, which will be governed by other factors.

Universities must look to prediction of trends such as these in planning for their futures, but no industry is as dependent on knowing how to aggregate as is the insurance industry. If teenagers are healthier than middle-aged folk but more reckless, insurance companies charge them less for health insurance or life insurance, but more for auto insurance. Actuarial rates, indeed, are all about how to structure populations and aggregate individuals into groups in order to make rates reflect comparative risks to the insurer. At times, the ways in which categories are formed are deemed socially unacceptable because they discriminate against certain groups; this consequence does not, however, undercut the basic logic. The paradox of insurance is that rates must be set and any rate structure is necessarily discriminatory because it assigns individuals to groups based on a few defining characteristics while ignoring the differences between individuals within a group. In regulating the insurance industry, society simply decides which forms of discrimination are acceptable, and which are not.

Because aggregation suppresses detail about variation within groups, different degrees of aggregation present very different views of system dynamics. Brian Arthur, one of the principal intellectual forces at the Santa Fe Institute, cites as a prime example the development of the videotape industry.[4] Videotaping was an innovation that could not be denied; independent of the technology that was developed, people were going to want the capability to make and view tapes that could capture anything from Baby's first steps to sports events and first-run movies. The dynamics of the aggregate market were driven by consumer demand for a product and by the industry's willingness to provide it for a price. Within the industry, however, a self-organizing dynamic drove events in classic nonlinear fashion: because of the incompatibility of different systems, industry's favorite way to frustrate consumers, the increasing popularity of VHS systems made Betamax equipment a foolish investment for most people.

There was no obvious overwhelming technological superiority of one system over another, but the inability of people with Betamax systems to play VHS tapes on their machines, or to have others view the tapes they had made, drove the Betamax system into oblivion. The competitive dynamics within the industry thus was governed by a very different set of factors than was the dynamics of the industry as a whole, though clearly there was interplay between the two scales.

The same phenomenon, not surprisingly, is played out over and over again in the commercial marketplace; equally compelling cases can be made regarding slide projectors, typewriter and computer keyboards, and even computer programming languages. Perhaps the best-known example involves the standard keyboard, which begins in the upper left with the letters QWERTY. Paul David argues that the current keyboard design is an accident of history: inferior even to some of the early alternatives, it has been locked in by historical accident.[5] Brian Arthur cites this and other examples, including the computer language FORTRAN, as cases in which technologies became established when competition was minimal and thereafter could not be dislodged, or were at least difficult to dislodge, even by superior designs.

A number of key ideas emerge from these examples, with broad implications for understanding evolution. For the introduction and diffusion of innovations, whether technological or biological, local *optimality* replaces global optimality as the order of the day. A solution that works can become established in the absence of intense competition, even if it is not optimal in any absolute sense. Once it is established, the competitive landscape becomes altered so that the sufficient solution can repel competitive forays by technologically superior solutions because of the dynamics of the marketplace or, in the case of evolution, the dynamics of the biological community. This is nonlinearity at its most evident: changes in mere abundance of a type can change the relative advantages or disadvantages of that type, sometimes favoring the further spread of that type, sometimes limiting it. Those changes in advantage may be enhanced and reinforced by further technological or evolutionary advances, such as the development and introduction of software and peripherals, or they may simply reflect the reinforcement of product loyalty on what would otherwise be a level playing field. Whatever the reasons, the adaptive landscape becomes a very rugged one, with multiple peaks and multiple possibilities for dynamics.[6]

François Jacob provides numerous examples of accidents of history becoming locked in through the nonlinearities of evolution and the process of tinkering.[7] A prime example is provided by the variety of different solutions that have arisen, in different groups of organisms, to the problem of vision. There are at least three different principles on which forms of eyes have arisen in evolution in response to the challenge of finding a way to see things: pinholes, multiple holes, and lenses such as are characteristic of vertebrate eyes. The evolution of eyes with lenses has been a complex process, dependent on a fortunate sequence of prior steps that allowed the development of lenses.

Adaptation builds on adaptation. As plants evolve to capture the sun's energy, pathogens and herbivores evolve to steal some of it; in some cases, associations can evolve further for the mutual benefit of both species. Some ant species, for example, specialize in particular tropical tree species.[8] The ants make their homes in acacia trees and in turn provide the trees some protection against other insect pests. This arrangement works out well for both species, and so the acacia has developed adaptations to make food more easily accessible for the ants.[9] Most important, these acacias, unlike closely related species, maintain leaves year-round, so that the ants always have a needed food source. Many other such mutualistic associations are to be found between species, as, for example, in the remarkable coadaptations of some plants and their pollinators, or the scaly *lichens* that unite species of fungi and algae.

These examples address evolutionary mechanisms and the question of why certain ecological relationships have arisen while others have not. Similarly, historical accidents reinforced by nonlinear processes account for differences among the biological communities to be found in quite comparable habitats. Chance colonizations result in *faunal* (animal) or *floral* (plant) differences among similar areas, as among the islands of Hawaii or the Galápagos.[10] These initial differences can so influence the development of the community that they affect the later success of other invaders. Thus, when rats were introduced as exotics into Fiji and Samoa, they had little effect on native bird communities because the existence of native rats or ratlike enemies (land crabs) of birds had "immunized" the community, selecting for behavioral repertoires that reduced mortality due to predation.[11] In contrast, similar exotic introductions into Hawaii and Midway, which had no native rat populations, had disastrous effects on bird populations, owing to predation.[12]

Nonlinearity means that one must examine evolution as a set of problems in game theory: a winning type is not necessarily the best of all solutions, judged against some absolute standard; rather, it is a type that, once established in the population, cannot be displaced. This is an obvious way of looking at things when thinking about economic competition, and it is equally valid for the natural world. It assumes that the broad outlines of evolutionary adaptation in the face of environmental change may be predictable, but that the details are not.[13] Even with hindsight, therefore, we are constantly amazed at some of the novelties that the evolutionary process has crafted. It explains in part the incredible diversity of ways that different organisms have evolved to meet similar challenges, such as the escape from unfavorable conditions, or simply the vicissitudes of their environments.[14]

The Broad Scale: The Zones of Holdridge

The primary determinants of vegetation are precipitation and temperature, which combine to govern the availability of water and nutrients. Polar environments are on the whole cold and dry; things get hotter and wetter as one moves toward the equator. There are exceptions, but given the way that the Earth moves about the sun, the broad features must be true. This gradient in temperature and moisture, as modified by topographic features such as mountains and valleys, provides the background against which plant life is distributed. A particularly useful complement to this rule is the Holdridge life zone system, presented by Leslie R. Holdridge in a classic paper in the journal *Science* a half-century ago.[15] The Holdridge scheme classifies vegetation into a variety of types—such as desert scrub, rain forest, and dry tundra—and relates these to average total annual precipitation and mean annual temperature. The Holdridge system is only one of a variety of efforts to relate vegetation to climatic conditions, but it serves as well as any. Together with the generalization mentioned earlier about latitudinal trends in precipitation and temperature, it offers some important insights into latitudinal trends in plant species diversity.

Within a broad latitudinal zone, there is, of course, much scope for variation. Philadelphia and Madrid both are at 40 degrees north latitude but experience very different climates. Traversing the globe, at 40 degrees north, one finds, depending on longitude, a wide range of vegetation types, from mountains to deserts, from Mediterranean scrub near Madrid to deciduous broadleaf forests near Philadelphia. Climate patterns vary a great deal, and their description is an essential refinement of the boundary conditions on vegetation types. Average conditions also do not suffice for determining vegetation type; seasonality matters, of course, a great deal.

Mesoscales, Communities, and Superorganisms

The delineation of species distributions in response to variation in environmental factors, at scales finer than those accommodated by the Holdridge system, often can be studied by careful description of the distribution of species along gradients in space. Places can be characterized by their patterns of temperature and precipitation, including how those critical variables fluctuate over time; those same sites then are also characterized according to their vegetation. From this coincident information, one can construct candidate schemes for relating vegetation to climate.

The most influential studies of this kind were carried out by my late and close colleague Robert Whittaker, along altitudinal gradients in habitats such as the Siskiyou Mountains in Oregon and the Great Smoky Mountains. I can still see Bob Whittaker, with his long strides and seven-league boots, gobbling up distances between home and office in Ithaca with the same purposeful mien that I saw the few times I was fortunate enough to be in the field with him. I can imagine him as he was in the Siskiyous and am hardly surprised that he was able to cover so much territory and achieve so much. In addition to chronicling a great deal of information about how individual tree species were distributed in relation to climatic conditions, Whittaker—through his data collection and analysis—produced the most telling evidence that has been collected against the then-popular notion that communities were like *superorganisms,* fixed in their composition and representing evolutionary units uniquely determined by local climate.

Early in this century, two elegant and competing views of plant communities were offered. The first, put forth primarily by Henry Gleason, argued that as one studied the changes in plant species composition along gradients—for example, up mountain slopes—individual species made their appearance or disappearance independently of other species. In this view, plant species were *individualistically* distributed, each according to its unique relationship to the environment. The alternative view, advanced by Frederic Clements, was that plant communities were units; in the Clementsian view, as one moved along a gradient, one would reach points, thresholds, where one group of species dropped out and was replaced by an essentially completely different assemblage. Thus, a community was like a superorganism, which responded to evolutionary forces as if it were a single unit. The Clementsian viewpoint was very attractive and much more popular than the Gleasonian view—that is, until Whittaker. Whittaker's data showed that, along the gradients he studied, species came and went by themselves, and that the notion of a superorganism was not valid.

Since the superorganism point of view raised the unit of selection to much higher organizational levels than were appropriate, the implications of Whittaker's work for the theory of evolution of communities and ecosystems were far-reaching. The issues were quite similar to those that surround the debate over Gaia mentioned earlier. That is, Clements viewed the ecological community as a superorganism; Gaia views the whole biosphere as such. In each case, the issue is whether important evolutionary forces can be acting at such aggregated levels, or whether we must understand system properties as simply having *emerged* from pressures at lower levels of organization. In the case of ecological communities, the problem is confounded by the fact that what we call a community or an ecosystem is often a fiction, an arbitrary restriction of spatial boundaries rather than a reflection of real thresholds of species change.

Whittaker's work does not imply that there are no thresholds, of course; along any gradient there may be major points of transition, or *ecotones:* places where community composition changes dramatically and in concert. The edge of a lake or ocean provides an obvious example: clearly the characteristic organisms of the deep ocean differ from those of the coasts, whose organisms in turn differ from those of the inland regions. But these boundaries distinguish large and diffusely related groups of species, not the small number of tightly bound species that could serve as essential evolutionary units.

Marine communities exhibit their own characteristic patterns of distribution, governed largely by ocean circulation and the distribution of nutrients and organisms. Across the globe, regions of high marine *primary production* can be found in polar and coastal areas, as well as in upwelling areas (such as at the equator) that bring up nutrients from the depths. Open ocean areas and the central oceanic gyres are typically regions of low productivity. Primary productivity refers to the conversion of solar energy into carbon compounds, and it fuels the base of the food chain. In the oceans, primary productivity is primarily the domain of the small plants, phytoplankton, that inhabit the upper layers. Thus, the distributional patterns of productivity refer essentially to the distributional patterns of the production of phytoplankton. The availability of energy in the form of phytoplankton, of course, determines where the small animals, the *zooplankton,* can grow, and so on to fish, although the high mobility of fish species gives them somewhat broader distributions.

Vertical patterns are also important within the marine environment, just as they are on land. Light can penetrate only so far into the water, giving rise to the description of the upper layer as the *photic zone.* Anyone who has ever scuba-dived knows that the depths are increasingly devoid of light, but that where things become dark depends a great deal on certain factors, such as the constitution of the bottom, that influence the clarity of the water. Water clarity above coral reefs is excellent to great depths compared with muddy lakes, which become dark a few feet from the surface. In any case, and in virtually any but the most shallow environments, the distributional patterns of species vary a great deal with depth. Environments are typically characterized by the rate at which light is extinguished with depth; this determines not only the widths of particular zones but also their degree of communication with one another. If light penetrates deeply, so that zones are broad, then the exchange between zones is relatively smaller in effect than if zones are narrow.

Summing Up

This chapter has shown that there are general rules that govern the kinds of plants and animals that one may expect to observe in different classes of environment. Those rules necessarily deal with broad classes of species, typically distinguished and identified by the adaptations that enable them to live in those environments. To a lesser extent, one may predict the appearance of particular species in particular habitats, but the reliability with which that can be done depends on the commonness of the species and the size of the area in question.

There are immediate and obvious economic parallels. Any large city has characteristic zones, perhaps associated with areas near large highways, where one can reliably expect to find the sorts of conveniences that travelers enjoy: places to fuel their automobiles and bodies quickly and to recycle unwanted wastes. Thus, service stations and fast-food establishments dot the highway exits, which are more numerous near cities. At any exit, one may hope to find an assortment of types of gasoline and types of food, but with no guarantee that one will find any one establishment in particular.

The fast-food establishments are a functional meta-group, composed of smaller functional groups specializing in burgers or fried chicken or pizza. Within any particular functional group, there is considerable redundancy, and types are largely indistinguishable except by the burger or pizza connoisseurs and cognoscenti. One is most likely, of course, to find McDonald's, the most prevalent species of its type, but that cannot be guaranteed. If one is not committed to a particular exit but willing to get off at any one of the "Next 17 Exits" that yield multiple points of entry into Gotham City, then more specific preferences can be indulged, since the increase of spatial scale increases the likelihood that a desired species, such as McDonald's or Wendy's, can be found. Thus, both the spatial scale of interest (the area considered) and the scale of functional resolution (whether one will settle for any fast-food restaurant or must have a McDonald's) affect the ability to predict what is likely to be found. In this particular example, temporal scale is less important, except to the degree that time translates into space as one circles the beltway. It really will not do, for example, to get off at exit 18 and wait until a McDonald's is built; the scales of waiting and building are simply not commensurate. In other situations, however, temporal scale can be very important to predictive ability.

For plants or animals, the scales of many fluctuations are rapid enough that strategies have simply evolved to average over them or to shut down activity during unfavorable times and gear it back up during better times. This is evident in the daily patterns, the so-called diurnal rhythms, that govern sleep-wake cycles, as well as in the yearly patterns of activity and hibernation, or of growth and dormancy, that represent the ways of life for many species.

In many ways, variability in space and in time is interchangeable. Small patches of forest, for example, are highly variable and highly unpredictable. Larger areas tend to be less variable and more predictable. That fact has not been lost on plant species in their evolutionary meanderings. Strategies have evolved to deal with unpredictability; greater predictability may be achieved by averaging over space, through dispersal, or over time, through dormancy or by becoming perennial. Space and time are two different ways for plants to assemble their mutual funds, reducing risk while eschewing potentially larger payoffs; evolution is for the long haul. Animals, of course, employ similar strategies in their foraging behaviors, searching broadly (as do highly mobile vertebrates) or sitting and waiting (as do many intertidal invertebrates).

Thus, on broad spatial scales and long temporal scales, ecological systems exhibit well-established patterns of species distributions, as reflected by schemes such as Holdridge's. These deal most reliably in generalities, and least reliably in the prediction of exact species composition. For many human purposes, however, that is just fine, since it is exactly the broad features that somewhat interchangeable species share that represent what is important to us. This is no accident; having alternatives in terms of resource species is an averaging strategy not so different from averaging over space and time. Specialists are ecologically vulnerable; generalists have options. Resiliency, which here translates into the ability of a species to survive even in the face of the loss of an important resource species, depends on flexibility, the ability to substitute.

In the financial markets, there are niches for all sorts of investment strategies. Most small investors are somewhat conservative generalists, developing averaging strategies to ensure a reasonable return while guarding against large losses. Others are willing to accept short-term setbacks for potential big payoffs and thus prefer more aggressive specialist strategies, focusing on a few risky investments. As one gets older, with a shorter remaining time horizon, conservative strategies become more attractive. There is a similar diversification of niches in ecological systems. Most species develop averaging strategies, for example, by wide distribution of their seeds, so that they are reasonably assured of some return. Most seeds will be wasted, but a few will find open gaps in which they can persist long enough to grow and reproduce. At the other extreme are species looking for the big payoff—a patch of land that they can settle into and dominate for many years. These are the *climax* species in the forest, the large trees, such as beech or maple, that fill the canopy and define the forests. In between the early transients and the climax species are a whole progression of others that are intermediate on the transient-climax spectrum.

In toto, this spectrum of species defines what is called a successional gradient; through the partitioning of this gradient, the system maintains its diversity. The classification into risk-taking or conservative species, however, is not so simple and is somewhat paradoxical. The species that average by distributing their seeds widely are generally termed *opportunistic* species, since they make their living by putting individual seeds into high-risk situations. A few seeds will find a rare and risky opportunity for high payoff, but most of them will be lost. The parent plant, however, is following a very conservative investment strategy, producing huge numbers of seeds that are individually very cheap to produce and sprinkling them over the landscape in the relative assurance that a few of them will strike paydirt. The large canopy species, on the other hand, produce smaller numbers of larger seeds, perhaps involving elaborate shells for protection and hence requiring a great degree of investment. Each seed is more carefully and conservatively invested, but there are far fewer of them.

The spectrum from opportunism to long-haul strategies has obvious parallels to economic markets. As new technologies emerge, the number of start-up companies proliferates from only a few to hundreds within a few years; many of these ultimately disappear, however, as a process of winnowing reduces the number of companies back to a few dominants. The same is true of a developing patch of forest undergoing *secondary succession*—that is, the reinvasion by a broadly predictable sequence of species following a disturbance. In the early stages of recolonization, there is very little competition and great scope for growth. The species that appear are those that are adapted to finding favorable places for fast growth. If the area that has been disturbed is large, there are lots of potential colonization sites, and lots of ways for a variety of species to enter the system. It is no surprise, then, that there is very high species diversity early in the successional process, or indeed that such diversity increases as the system develops. At some point, however, the honeymoon is over. The growth of these thousand points of flight begins to bring beachheads into contact with one another, and competition becomes increasingly important.

The same scenario is played out dramatically on coral reefs and has led to the evolution of a variety of aggressive ways for one coral species to engage in antisocial behavior, such as poisoning competitors. The corals are the Borgias of the invertebrate world.

The increased importance of competition for space in the later stages of community development leads to a winnowing of species and a decrease in diversity. In the absence of new disturbance, the forest community will move inexorably toward dominance by one or two world-champion competitors. Disturbance is a constant feature of natural forests, however, at a variety of scales, so in a typical large forest diversity is maintained by the continual playing out of the disturbance-recolonization dynamic on smaller scales. Recurrent disturbance, especially forest fires, plays a vital role in maintaining diversity and resiliency in the forest, as well as in most other natural systems, serving to stimulate a process of renewal. This somewhat counterintuitive finding, to which I return in the next chapter, has changed the way forest managers deal with fire suppression.

Though exact species composition is unpredictable, descriptors of biodiversity typically follow more regular rules, since these are statistical compressions of a great deal of information into a summary measure or set of measures. Just as averaging over space or time can lead to more predictability, so, too, can averaging over details of the exact composition of an ecosystem. Biodiversity does not refer to a single number, though discussion of it usually begins with species counts.

Equally important, however, is the diversity of populations within a species, the diversity of species within a functional group, and the diversity of functional groups within an ecosystem. Furthermore, species themselves are not monolithic collections of indistinguishable units; genetic and demographic diversity within populations plays a vital role in establishing the resiliency of ecosystems and their ability to continue to provide the goods and services we rely on.

Thus, biodiversity is distributed, like the stars and gases and dust in the sky, into nebulae and galaxies of determinate and indeterminate structure, producing a picture whose intricacies defy simple description. Any effort to measure biodiversity collapses this rich tapestry into simplifications that we can fit into our brains, our models, and our laws. That does not diminish the importance of settling on suitable ways of describing biodiversity that capture its essential importance to the human endeavor, while suppressing much of its glorious detail. The issue is to understand which detail is important to the maintenance of services, and which can be suppressed without changing the basic message. Counting species is not the answer, because we know that species differ a great deal in their importance, and even weighted species counts would miss the critical questions of how biodiversity is organized above and below the level of the species. Deciding on the relative importance of different components of biodiversity must depend on an understanding of how systems are organized and of how structure and functioning are related. It is a question, like most, more easily posed than answered.

Measurement of biodiversity, and of its importance, must rise above simply providing lists of parts and transcend a taxonomy, however good, that simply arranges things within boxes. One must, as John Holland's list of four key properties emphasizes, also talk about flows and nonlinearities—that is, statics and dynamics. Flows are easier to describe than nonlinear dynamics, because it is feasible to measure them by observation alone, and over relatively shorter periods of time. Economists measure the flows of goods and services through an economy, sociologists and cultural anthropologists measure the flows of customs and information, and toxicologists and epidemiologists measure the flows of harmful agents. The nonlinear consequences of those flows are another matter, potentially of far greater importance depending on how buffered the receiving systems are against change.

Ecologists similarly describe systems first of all by the flow of materials through compartments, in the same way that physicians or physiologists describe the flow of blood or nutrients or coffee through an organism, or economists the flow of capital through an economy. A farm ecologist describing the flow and *residence times* of nitrogen in an agricultural ecosystem, from fertilizer to crops, is thus engaged in an activity analogous to what a pharmacologist does in describing the flow and residence times of drugs in the body, going from one organ to the next. The ecological question, as Joel Cohen put it so impeccably, is, "Who eats whom?" The process of gathering such culinary data is not typically very appetizing, often involving exercises in vivisection and analysis of the stomach contents of large numbers of individuals. But it's a job, and someone has to do it. From such studies we have learned a great deal about the *food webs* (more technically termed *trophic webs,* from the Greek word for food) of a variety of ecological systems. From inventories of food webs, such as those organized by Frederic Briand and Joel Cohen, and insightful analyses one can begin to discern patterns, commonalities, and differences among the organizational structures of diverse food webs.

It is has long been well understood, from purely thermodynamic considerations, that there are limits to the number of layers food webs can have, that is, to the number of steps energy can take in going from the sun to the top of the chain. At each step, typically at least 90 percent of the energy is lost in conversion from the biomass of one species into another. Thus, a classic pyramid structure forms, in which there is far less energy available at each trophic level than at the level on which it feeds. The notion of trophic level is an oversimplification for the convenience of discussion. Ecosystems are not organized into layers, each feeding only on the layer below; lines are crossed freely. Humans, for example, eat pretty much anything. Nonetheless, the basic message remains that any parcel of energy can make just so many stops on its journey, since it becomes dissipated at each step along the way. Given that such limits exist, we need to ask whether they are reached in practice, or whether other constraints—for example, ecosystem stability—impose stricter limits. The basic issue, the degree to which food web dimensions are extrinsically determined versus the degree to which they are intrinsically determined, has been the focus of clever analyses by a variety of scientists. A valuable synthesis is provided by Stuart Pimm in his book *Food Webs.*[16]

Not just the length of food webs but their topologies as well have captured the imaginations of investigators. Common patterns, such as the relative numbers of species at different trophic levels, have been detected, most dramatically in the work of Joel Cohen,[17] although many of these patterns have recently been contested by other researchers.[18] Whatever the truth, it is clear that there are some commonalities to be found, especially in ecosystems of similar type.

More generally, as one looks beyond trophic webs to the comparative functional organization of diverse ecosystems that exist under similar environmental conditions, one finds similarities that reflect what biologists call *convergent evolution.* Convergent evolution refers to the tendency of separate organisms or communities of organisms independently to evolve similar adaptations and patterns in response to similar environmental challenges, though possibly separated by great geographical distances. A case in point is revealed by comparisons of the structure of ecosystems across a range of arid and semiarid regions.[19] Desert shrubs, for example, typically have succulent stems wherever they are found. Thus, succulent stems typify these plants. Similarly, Mediterranean-climate chaparral are *sclerophylls* (have hard leaves); thus, sclerophylly is a common characteristic of chaparral plants across the globe.[20] More generally, plant morphology can be predicted in general terms given information about the physical characteristics of the habitat.[21] Not only the life forms of individual plants but also community characteristics, such as productivity, food chain length, and broad features of patterns of energy flow, are determined primarily by the local physical environment rather than by evolutionary history. Thus, ecosystems in similar environments converge to resemble each other in their overall features, though they differ a great deal in specific detail.

In Conclusion

Through a bit of ecological detective work, we have learned that community composition and organization are the children of two parents: local environmental conditions, such as temperature and humidity, and the vagaries of history. Local conditions determine where forests can occur, and even the types of forests, but the particular identities of the species to be found are governed by chance colonization events and by what is available in nearby communities. Broader spatial areas are more easily predictable than small patches of forest, because chance events tend to be averaged out more. And what is true for the forest is true for the grassland, or for the intertidal zone of rocky coasts, or for wetland communities: there is predictability, and there is uncertainty, and it is that uncertainty—that unpredictability in local outcomes—that gives us the rich tapestry of diversity that can be seen in natural communities. Because of the uncertainty in outcome, each new gap that forms will follow its own unique pathway of growth and replacement, and the forest mosaic—or for that matter, the intertidal mosaic, or the grassland mosaic—will be a patchwork of accidents of colonization and later growth. Uncertainty, thus, is the parent of biodiversity, and localized disturbances such as fires—which, at least up to a point, increase the level of uncertainty—are key to the maintenance of diversity.

Ecosystems are self-organizing systems in which random disturbance and colonization events create a heterogeneous landscape of diverse species, which then become knitted together through nutrient fluxes and other forms of interaction. Not all is chance, however, as generalizations of the sort discussed in the last section show. Ecosystem development becomes channeled into one of a limited number of typical patterns, determined by local environmental conditions as well as by initial events early in their *ontogeny* (their development). If we were to classify each system type in terms of a basic organizational plan, we would have a powerful tool in developing a general theory of ecosystem structure and functioning. What do general theories of complex adaptive systems have to tell us about the commonalities to be expected, and the roles of extrinsic and historical factors? I turn to these questions in the subsequent chapters.

5

ECOLOGICAL ASSEMBLY

The assembly of an ecological community following a disturbance, a process called *succession,* is quite different from putting together a new bookshelf from its parts; indeed, it is more like playing with a set of Lego bricks. There is no unique blueprint, though there are a number of basic forms that can be created. There is considerable flexibility in design, but there is also a great deal of historical constraint, in that the way the first pieces are fitted together limits the possibilities to follow. Tinkering, again, is the dominant metaphor and leads to a multiplicity of possible combinations and structures. In general, not all pieces will be used. Among those that are used, some will serve as keystones: their removal would lead to the collapse of the integrity of the structure. Others will be adornments: their removal would have minor effect.

Where ecosystems differ from most Lego structures is in the constant turnover of the pieces, at least at local scales. Species come and go, replaced by others that appear to take their place. This has been most dramatically demonstrated in the study of islands, which have provided laboratories for ecological and evolutionary studies since Charles Darwin's Galápagos voyage on the *Beagle.* Precisely because the assembly process has so many twists and turns, groups of similar islands provide naturalists with the opportunity to observe the colonization experiment many times over. Any species differences between islands may in part be due to underlying habitat variation, but they may also be by-products of the vagaries of the assembly process, through which accidents of colonization are transformed through subsequent events into facts on the ground, constraints on the further development of the island biota. Sorting out the relative importance of extrinsic and intrinsic factors, and of fate and chance, thus looms as the essential issue in the comparison of the biology of similar islands. This now-familiar challenge, which has arisen over and over again in this book, is also a central problem in the study of complex adaptive systems in general.

The Biology of Islands

Following Darwin's original fundamental work, numerous other investigators carried out biological studies of islands. In general, interest was focused on how species composition could be related to the specific characteristics of individual islands. Sherwin Carlquist, in his book *Island Life,* summarized much of what was known in 1965, which was a whole lot less than we know today.[1] At about that time, a unique combination of three scientific talents—Robert MacArthur, Edward Wilson, and Daniel Simberloff—realized that a number of things had to be done if the anecdotal information that had accumulated was to be shaped into a science of islands.

Robert MacArthur, at Princeton, was undoubtedly one of the most creative and important influences on the transformation of the science of ecology into its modern form. His roots in the natural history of birds and his ability to crystallize observations into neat and simple mathematical packages pointed the way to generalizations that provided evolutionary context for understanding why island biotas have the species that they do. His crisp and elegant writings over too short a career inspired a whole generation of young ecologists, while providing enough rigor to ecological relationships to attract the attention of researchers from outside the discipline and introduce them to the fascinating puzzles of evolutionary ecology. Many of his disciples had been originally trained in mathematics or physics, and they would become leaders in the development of theoretical ecology in the decades that followed MacArthur's untimely death at the age of forty-three.

Like Lawrence Slobodkin, MacArthur was a student of G. Evelyn Hutchinson, whose role as a mentor of brilliant young ecologists remains unsurpassed. An incisive theoretician as well as a stunning empiricist, Hutchinson brought both these talents to bear on his writings, whose substance and imagery captured the reader's attention and impelled him or her to want to study the subject further. His style of training graduate students, as Larry Slobodkin once explained to me, was to encourage, reinforce, and get out of the way. He also chose his students well, and his incredible legacy is testimony to the success of his philosophy of mentoring.

MacArthur was a worthy intellectual descendant of Hutchinson's, who nonetheless outlived his student by two decades. Though Hutchinson was not afraid of mathematics and used it insightfully to support his writings, MacArthur made greater use of it as a vehicle for excursions into the fantasy world of what might be and what might have been. It was important, therefore, that he found an empirically based collaborator of equal stature in Edward O. Wilson, who joined him in his most lasting and influential work, *The Theory of Island Biogeography.*[2] Wilson, revered among myrmecologists (ant fanatics) for his landmark studies of the social behavior of ants, is more widely known for his stimulating and at times controversial writings on human sociobiology. He remains to this day one of the most important biologists in the world.

The combination of MacArthur and Wilson was a collaboration made in heaven, bringing together two of the true geniuses in the subject. Together they crafted the fundamentals of a theory of *island biogeography,* a theory that endeavored to explain the diverse biotas of groups of islands in terms of such properties as size, topographic complexity, and distance from the mainland; they recognized as well that accidents of colonization and extinction were fundamental determinants of the exact species composition, but nonetheless they sought to derive regularities from a sea of detail.

The most profound prediction of the MacArthur-Wilson theory was that islands are constantly in flux in terms of their species composition, but that they reach a balance between arrivals and departures so that the number of species reaches an equilibrium. The idea of such a dynamic equilibrium is similar to what happens if water is run into a bathtub in which the plug is missing. If the inflow rate is strong enough, initially water flows in much faster than it flows out. As the amount in the tub increases, the outflow rate also increases until it reaches a level equal to the inflow rate. When inflow equals outflow, there is no change in water level, although there is constant turnover of the water. Such turnover is also characteristic of lakes, which are just big bathtubs with the plugs missing. Water is constantly flowing out of lakes, into rivers and streams and on to the oceans; it is replaced by new water, from snowmelt and rainfall at higher elevations. Lake scientists, who are called *limnologists,* characterize a lake by its *flushing rate,* basically the rate at which the water in the lake is renewed. Since the flushing rate determines how long a packet of water will remain in the lake, it also determines how long toxic materials that have entered the lake will remain there—that is, what their residence times will be.

MacArthur and Wilson thought about islands as if they were bathtubs. Species flow into an island via immigration; they flow out by local extinction. An empty island should have a relatively high inflow rate because every mainland species is a potential colonist. As the number of species that have been successfully established on the island increases, the number of potential colonists decreases. Furthermore, the mainland species that have not yet managed to colonize are likely to be the less vagile—that is, less likely to be able to travel from the mainland to the island. Insects that attach themselves to birds, for example, can in theory make their way quickly to islands; those that must cover the distance themselves, or attach to flotsam, might delay their appearance. Thus, although there may be exceptions to this rule, MacArthur and Wilson hypothesized that immigration rates should decrease as the island biota developed. On the other hand, extinction rates should increase as the island becomes more crowded, both because there are more candidate species for extinction and because competitive interactions increase as species expand and come more and more into contact with one another. Such competitive interactions lead to the eventual elimination on the islands of the inferior competitors, in the classical Darwinian manner. Eventually, the number of species should reach an equilibrium level, a dynamic balance between the two processes of immigration and extinction. Instead of a flushing rate, island biologists speak of a species *turnover rate,* but the concept is the same. Species come and go on the island. Some stay virtually forever, or at least until a major catastrophe such as a hurricane hits the island; others have quite a short residence time.

MacArthur and Wilson made a number of additional predictions, all potentially testable. Whatever the immigration and extinction rates, some equilibrium should be reached representing a balance between them. Islands far from shore should have lower immigration rates, and hence fewer species at equilibrium. Similarly, smaller islands should have higher extinction rates (per species), again leading to fewer species.

With this simple and elegant theory in hand, they set out to find a way to test it. Graduate students are always useful in this regard, and Wilson found a particularly bright and industrious one in Daniel Simberloff, now a professor at the University of Tennessee. For his thesis work, Simberloff (after tortured consideration of the ethical issues involved) fumigated small mangrove islands in the Florida Keys and carefully studied the recolonization process. The essential predictions of the theory were borne out, and Simberloff's work remains a classic study today.[3] It was the experimental verification of MacArthur and Wilson's ideas that transformed armchair speculation into science, and into a fundamental element of current ecological theory and practice.

Controlled experiments such as Simberloff's are invaluable, but rare. The costs are large, and the moral implications always are present. Mangrove islands, populated mainly by vegetation and insects, are one thing; it would be virtually impossible to reproduce such experiments where large vertebrates were involved. A number of natural experiments do exist, however, and have provided targets of opportunity for colonization biologists. Classical examples include the recolonization of Krakatoa, an island west of Java, following a volcanic eruption that destroyed it more than a century ago. Although it represents a unique, nonrepeatable occurrence, Krakatoa still provides a long-term laboratory for studying the process of *primary succession.* Primary succession refers to the colonization of a new substrate, in this case, the lava that covered what was left of the old island. During primary succession, there are important physical and chemical changes to the substrate that result from the activity of the species that have colonized there. In this way, primary succession differs from secondary succession, which refers to the recolonization of an area where the local plants have been removed but the basic physical properties have not been so disturbed as to eliminate the effects of earlier primary succession.

As would be expected from general theory, the number of plant species on Krakatoa continually increased over time and was still in exponential increase when Willem M. Doctors van Leeuwen published his study of the recovery process.[4] Mosses and lichens colonized first, followed by more complex species such as flowering plants. Similar studies, with similar results, have been carried out for other sites, such as the new island of Surtsey, which formed as the result of a volcanic eruption in Iceland in 1965, and the areas near Washington State's Mount Saint Helens, which blew off its scenic top just a few years ago.

The Mount Saint Helens studies, carried out largely by terrestrial ecologists such as Jerry Franklin and Peter Kareiva at the University of Washington, emphasize the importance of the theory of island biogeography to islands of habitat in all sorts of systems, including forests and grasslands. Thus, the ideas have relevance well beyond the conventional sorts of islands that Simberloff, MacArthur, and Wilson studied or discussed—bits of land surrounded by expanses of saltwater. MacArthur and Wilson, of course, knew that they were developing theories of deep significance for a wide range of systems, not just for mangroves. This was, in fact, their motivation; neither had paid much attention to islands before. As fascinating and romantic as islands are, their greater value is as model systems. In this, they play a role very similar to mathematical or scale models—they simplify interactions, emphasizing and exaggerating a few dominant processes at the expense of the complexity and detail that one finds in more enmeshed situations. Thus, even large islands such as Australia, New Zealand, and Hawaii provide ideal systems for study. Their distance from other sources of species has isolated them in their evolutionary development, allowing the establishment of unique types of plants and animals—for example, the marsupials; on these islands, researchers can focus their attention on the same basic mechanisms that form the cornerstones of the MacArthur-Wilson theory. Australia is large, but far from any other source. New Zealand is smaller, and even more isolated. The Hawaiian Islands form a tiny chain, a dot in the middle of the Pacific, far from any mainland source. The MacArthur-Wilson theory makes strong predictions about the comparative dynamics of these different land masses, and those predictions are consistent with what is to be found there. Hawaii provides an object lesson in how to measure distances, however; humans have introduced large numbers of species into Hawaii, in effect reducing the distance of the islands from the mainland. The biota of Hawaii is a mess, dominated by introduced species to the extent that very few native species remain.

The lessons learned from island models are especially valuable in studying virtual islands in terrestrial situations—for example, patches of natural area in the midst of urban areas, or gaps caused by natural or human influences in forests. Indeed, forests are mosaics of gaps in various stages of recovery from disturbance and thereby support a diversity of different plant types. The process of recovery from the natural deaths of trees in a forest is fundamentally different from what happens following a volcanic disturbance; instead of bare rock, for example, the substrate is developed soil. Hence, the process is termed secondary succession, to distinguish it from the more fundamental primary succession. The lessons from island biogeography provide great insight into what will happen to a forest patch. Its size, its complexity, and its distance from source areas will all influence its development.

Islands in the Forest

The theory of island biogeography, suitably extended, is central to understanding the dynamics of forests. In forests, gaps are constantly forming as the result of the natural deaths of individual trees (or simply the loss of branches), the blowdowns of large numbers of trees during storms, or the demise of whole areas due to forest fires or clear-cutting practice. The source of the disturbance affects the recovery process of any such gap in the forest: it determines the quality of the substrate and whether there are seeds remaining from which to rebuild. Also important, however, as in the basic theory of island biogeography, is the size of the island gaps and the degree of isolation from sources of colonists.

In most forests, localized disturbances are a vital part of their continual renewal; small fires, for example, once the clarion call for Smokey the Bear to come to the rescue, are now recognized as essential for maintaining the health of the ecosystem. Fire opens up space where species that cannot tolerate shade—like larch and various pines in the western mountains of the United States—have the opportunity to flourish. It contributes to the recycling of nutrients. And most important, it introduces diversity and heterogeneity to the forest, helping to "immunize" it against more major disturbances, including especially catastrophic fires that might destroy the entire forest. In a healthy forest, little disturbances—deriving not only from fire but also from natural mortality, windthrows, and disease—occur sporadically all over the landscape, reinitiating local sequences of secondary succession. Certain species, such as mosses, herbs, and shrubs, are best adapted to the early stages of the successional process; others, including small conifers, do best in the intermediate stages; and still others, such as hardwoods, thrive during the late stages. In Connecticut forests, for example, black cherry and red oak make their appearance fairly early following a disturbance, hemlock comes in later, and beech is adapted to the late stages of succession. The early species have made adaptations, such as developing high respiration rates, that permit them to perform well under high light; these species are hampered, however, by the cost of such adaptations, which become apparent as the forest develops: the canopy closes up and light becomes less available.

The Ecological Quilting Bee

The typical pattern of development varies from system to system, but it is the sharing of different stages of successional development that maintains diversity in virtually any ecosystem. The natural forest becomes a tapestry of patches in different stages of succession, and hence a tapestry of diversity. The small local disturbances not only maintain the character of the system by maintaining the species that are early colonists but poor competitors; they also maintain the *resiliency* of the system, preserving the opportunistic species that thrive under the conditions accompanying the unpredictable but inevitable environmental changes that occur at broader spatial scales, such as massive windthrows or fire.

The lessons of this example extend beyond forests; they apply immediately to other vegetation systems: to grasslands, where small mammals such as pocket gophers may serve as the disturbing agents; to marine and aquatic systems, such as *planktonic* communities;[5] and to the rocky intertidal coasts. Indeed, it was in these glorious intertidal communities, in work with Robert Paine, that my understanding of such processes first came into focus.

The intertidal communities of the exposed Outer Coast of Washington State are dominated by a large bivalve, the mussel *Mytilus californianus.* This species is like a canopy tree in the forest: it dominates all other petitioners in the lottery for space. Just as the canopy species would turn forests into monocultures were it not for disturbance, so, too, would the mussels take over the exposed intertidal. Disturbance, however, dressed in various forms, removes the mussels, reinitiating a local successional process that maintains a wide variety of algae and invertebrates that would otherwise be little in evidence in the zones preferred by the mussels. The phenomenon is a vitally important one in many kinds of ecosystems.

But the news is not all good. The forces that maintain diversity by presenting havens for fugitives preserve good and bad alike. Pest species in agricultural systems survive on individual fields, from which they disperse to other fields before they can be exterminated. Infectious diseases of humans play a similar game, finding new patches of susceptible individuals that serve as a base for further spread before local health professionals can get an outbreak under control. Indeed, even a single host individual can be thought of as an island, a potential home for many different infectious agents. Each individual harbors many pathogens and parasites at any one time, so each is, in effect, an island occupied by many species. As the body rids itself of some parasites, it acquires new ones; a healthy individual is roughly in dynamic balance, at least over long periods of time. As with oceanic islands, the rate of acquisition of new parasite types by an individual depends on his or her degree of isolation or, in effect, distance from source areas. Indeed, just as for islands in the ocean, it may be that the number of occupying species gradually increases over time. The analogy breaks down, however, because increasing parasite burden (even for a single parasite species) can lead to the demise of the host individual; oceanic islands do not die because of an accumulation of species or individuals. This difference has important evolutionary implications for the coevolution of hosts and parasites, a topic to which I return in later chapters.

In large islands or systems of islands, new types also arise by *speciation,* the process by which existing species give rise to new ones.

There are also strong analogies here in the dynamics of infectious diseases. In particular, one hotly debated theory for how AIDS eventually leads to the death of an infected individual rests on just such a process. In this view, mutation of HIV within a host produces new types in a process analogous to speciation.[6] Eventually variants of the virus arise that overcome the immune system's defenses. The "diversity threshold" theory of AIDS progression may in fact turn out not to be correct, but the perspective that it adopts—of an individual host organism as a small ecosystem, populated by beneficial and harmful organisms in a struggle for hegemony—is a potentially powerful one for the management of disease.

Species-Area Relationships

Island biogeography predicts that, for a variety of reasons, larger areas hold more species. The reason is simple: species have more places to establish and persist in large areas, which in turn also provide a bigger target for colonists. Larger areas may also be expected to support a greater diversity of habitat types. At small scales, this heterogeneity favors, say, distinct forest species; at larger scales, the mosaic may include forests or swamps or other communities. Again, increased heterogeneity leads to higher levels of diversity. Indeed, the increase in species diversity from smaller to larger areas produces one of the most robust and well-established laws in ecology: as the area studied increases, the number of species also increases, as a power of that area.

This power law is not a law in the same sense as a geometric law or a law of physics; when we refer to laws in ecology, we generally mean that there are statistical regularities that seem robust over a wide range of examples. In geometry, we would say that the area of a square increases as a power—in particular, the square—of the length of one side. Or that the volume of a cube increases as a power—namely, the cube—of the length of one edge. We would know that this is exactly true for every square or cube imaginable.

Even in geometry, powers need not be nice integers, like two or three. We could turn around the above arguments, for example, and say that the length of the side of a square increases as the one-half power of its area, or that the length of the edge of a cube increases as the one-third power of its volume. For species on islands or in patches of different sizes within a single large area, the number of species increases as somewhere between the 0.1 power of area and the 0.4 power of area. The exact number varies for different groups of islands or for different types of habitats. Such "laws" thus are really rules of thumb, statistical regularities that admit many exceptions and variations.

Why do such regularities exist at all? Do they represent something unique to the number of species on islands? Or are they the natural consequences of a set of local rules of interaction that are common to a whole class of complex adaptive systems, of which species on islands provide only one example? If the latter interpretation is correct, then the regularities should apply more broadly. They should hold for the numbers of pathogens in diverse agricultural areas, or for the numbers of viral diseases prevalent in various cities. Larger communities should have more health food stores, sustain a greater diversity of political preferences, and even be richer storehouses of jokes. And in each case, there should be a nice power law relationship between whatever is being counted and the size of the region under consideration.

It may seem that I am comparing apples and oranges here, trying to find commonalities among a litany of quite diverse situations. There are, however, some common threads that link them together. All of the examples deal with comparisons among areas, and each area may be thought of as comprising lots of sites that are either occupied or not by each of the types being considered. For species on islands, the notion of *occupied* is pretty clear, as it is for most of the other examples. For jokes in a community, *occupied* means only that the people at a site know the joke.

Each example also describes a dynamic process in which the number of occupied sites changes over time. Geographical spread depends in part on an infectious process, generally with a strong local component: people tell jokes to their neighbors, give them horrible diseases, or persuade them to adopt or change their political positions. Health food shops raise public awareness of fitness and spin off branch stores that market the same goods. And so on.

Finally, there is in every one of these examples a process of local extinction through which diseases disappear (sometimes by killing their hosts), stores fail, and people forget jokes or decide that they are no

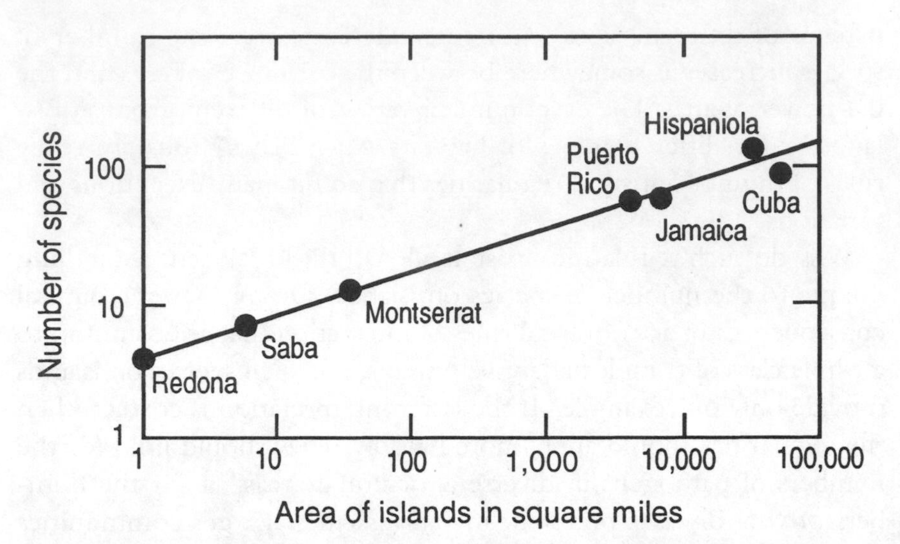

FIGURE 5.1 Species-Area Relationship for the West Indies. Illustration by Amy Bordvik.

longer interested in supporting a particular political candidate or position.

Thus, all of these examples have the same basic structure: a large area made up of many sites, each occupied or not by the various types. Local spread. Extinction. That is all. There is really nothing else going on. Hence, it is not surprising that models of all of these situations have similar features and give rise to similar patterns. The exact conclusions differ, of course, from situation to situation, because the details—for example, the methods of transmission and the levels at which these occur—are different. There is enough commonality among these examples, however, to argue for exploration of the simplest possible models that contain these ingredients, to see what generalities emerge.

This illuminates my philosophy of using models: start simple so that you understand what you are doing before moving on. Although it is important not to stop with these but eventually to introduce the detail that distinguishes one situation from another, simple models are a good place to start because their transparent features provide clarity. In the next several sections, I will introduce a framework for developing such models and then return to apply them to ecosystems in order to help us understand how observed species-area relationships arise.

Stochastic Cellular Automata

A central goal of scientific inquiry is to relate pattern and process—that is, statics and dynamics. The fundamental dilemma is that snapshots in time provide very limited insights into complex phenomena. Just as many roads lead to Rome, many dynamic pathways can produce similar patterns in any natural system. Mathematical and computer models can help us to make the connections between our observations and the mechanisms responsible for them. By exploring the outcomes or consequences of certain assumptions, these models allow us to produce a catalog of possible explanations; experiments and further observations then help in winnowing the list to establish probable cause.

A simple model is something to build on. In its sleek lines and limited assumptions, it can provide a base for elaboration while capturing the essence of a variety of more detailed possible explanations. A particularly useful platform of this sort is provided by a family of models known as *stochastic cellular automata*, or *interacting particle systems*. My frequent collaborator Richard Durrett, of Cornell University, is one of the gurus of interacting particle systems. He and others have demonstrated over a broad range of examples the remarkable power of such models to illuminate diverse phenomena.

Interacting particle models regard space as a grid divided into boxes or "cells," each in one of a finite number of conditions, termed states. In such models, individual cells change states according to sets of probabilistic rules, under the influence of their neighbors. The most basic examples of examples are called, evocatively, the *voter model* and the *contact process*. I will focus on the relevance of these models for ecological systems, but they apply equally well to other examples—such as the prevalence of businesses or political preferences, as discussed in the previous section.

The Voter Model

The voter model, as the term suggests, describes the dynamics of change in voter preference. It imagines a world in which no one has any fundamental convictions that affect voting behavior. Given a choice among candidates, each person arbitrarily picks one to support, but without feeling any real commitment. Feeling unsure, the

FIGURE 5.2
A snapshot of the contact process. Simulation by Linda Buttel.

individual voter constantly polls neighbors to find out whom they support. If the weight of opinion is against the current choice, the voter, with a certain probability, switches to the local favorite. Eventually, through this process, consensus develops, a consensus driven entirely by random events rather than substance. One can introduce more reality by assigning to individuals intrinsic preferences that provide resistance to the weight of local opinion; it is important, however, that we understand the simplest model before considering this complication.

The basic voter model provides a point of departure, a baseline for comparison. Each individual is assigned to a permanent location in space—that is, to a cell. A neighborhood size—say, the nearest four cells—is also specified. As one would expect, changing the size of the neighborhood—that is, changing the size of the peer group—changes the dynamics; indeed, if the neighborhood size becomes too large, the unique aspects of local interactions are lost. Television popularity polls serve the function of increasing the size of the peer group many times over, and as a result, a much higher degree of volatility is probably to be found in contemporary national election campaigns than was the case in years past. Changing the neighborhood size can also affect the final outcome. For example, for the voter model, whatever the neighborhood size, a consensus eventually forms in which all individuals think alike, but the winner may be changed, for a given initial distribution of preferences, by a change in the size of neighborhoods.

The basic voter model is not the end of our line of inquiry, but rather a foundation for refinements. We can learn more from the model if we imagine that the space in question is not actual physical space but rather an abstract spectrum of individual characteristics, such as socioeconomic status. Income level, for example, stratifies society into groups of people who interact more with one another than across levels, and who therefore influence each other's opinions more than they do those from other income classes. For some issues, there are intrinsic voting preferences associated with particular income levels, and the results of elections then represent the interplay between such preferences and local persuasion by neighbors. Consensus may form *within* individual income levels without necessarily reaching across levels. Global consensus thus may not be achieved, and majority rule or some surrogate must be invoked to effect decisions.

In other cases, however, people within particular income groups or different political parties may have no reasons other than custom and history for the specific positions they take on selected issues; that is, the attitudes taken by individual groups are associated with accidents of history and probably are not very deeply held. In this volatile situation, dramatic shifts in positions can occur, and consensus can develop. Given the controlling role of money in any economy, it may seem unlikely that positions on issues would not bear at least some essential relationship to wealth, but evidence to the contrary may be found in the apparent paradoxes in commonly held stances—say, on individual rights and liberties, affirmative action, or single-sex schools. My argument here, surely debatable, is that the traditional litmus tests of most sociopolitical agendas necessarily contain internal contradictions. Quotas in hiring, for example, reflect the best and worst of human intentions—the antipodal motivations to be inclusive and to be exclusive. Similarly, single-sex schools represent efforts not only to deny women equal access but to provide them with the best of opportunities. For such cases, contradictions seem unavoidable, and congruence with the fundamental philosophical agendas of particular political groups is less than clear. Positions on many such issues thus may have been codified as parts of the platforms of particular groups owing to accidents of history, and they may simply have become accepted by members of those groups as the dues of association. This is the voter model at work.

The Contact Process

The contact process is an extension of the voter model in which an individual may decide not to vote for anyone at all. More precisely, think of a landscape as made up of sites that are either occupied or not; an occupied site may be held by one of any number of possible types. Occupied sites may represent voters with a particular preference, and unoccupied sites those who have no preference. It is perhaps more natural, however, to think of the contact process in regard to ecosystems, in which types represent different species. It is, indeed, an ideal vehicle for starting to think about the mechanisms determining biological diversity.

Only two kinds of events are permitted: an occupied site may become extinct, or it may send out colonists to a neighboring unoccupied site. Each of these events is allowed to occur with a certain probability per unit of time, and the result is an ever-changing landscape that roughly resembles a forest or grassland, or perhaps the geographical distribution of pathogens. As each type becomes established locally, it spreads to neighboring areas; the consequence is the rapid development of clusters of individuals of particular types. Over longer time scales, those clusters interact with one another, giving rise to a slower dynamic that controls the overall diversity of types.

Variants on the contact model are very useful tools for investigating mechanisms of colonization and coexistence—the very factors that help generate and maintain biodiversity. Through such extensions, we can explore how basic details of the local system—the size of a neighborhood, the frequency with which sites are disturbed (the *death rate* of cells), the rate of colonization of neighboring sites (the *birth rate*), and dispersal distances—affect its characteristic features or those of any ecosystem. Typically, such features would include the abundances of various species in relation to the area studied and the fluctuations of those abundances—the most basic measures of the dynamics of biodiversity. Simple extensions of such models have proved useful in conservation biology. Roland Lamberson, Barry Noon, and their colleagues, for example, implemented very similar models to study the effects of logging practice on habitat for the spotted owl, helping to inform debate on the preservation of old-growth habitats in the northwestern United States.[7]

With my associates Kirk Moloney and Jianguo Wu, I have used elaborations of models just like this one to simulate the behavior of annual grass species in the serpentine soils at Jasper Ridge Reserve, above Stanford University; in this effort we have benefited from decades of research by the distinguished plant biologist Harold Mooney and his colleagues. Chocolate chip cookies, which my wife Carole baked and brought along, helped keep Hal Mooney involved, though I am sure the intellectual nourishment would have been sufficient. In the models we built, the Jasper Ridge landscape is divided into cells, each of which may or may not include a population, say, of the annual plant species *Bromus,* or of its competitor *Plantago,* or of some other species. The power of such models is that we can (in the computer) change small features of the system, such as the abundance of pocket gophers (which, by their burrowing behavior, create a force for disturbance similar to the role that fire plays in the forest), and in a matter of minutes explore the potential consequences for the abundance of the various plant species that exist in the grassland. Similar models have found wide application in a variety of systems, including especially forests, grasslands, and intertidal zone mussel beds.

The most rudimentary models of this kind must be enhanced with detail if one wants to make predictions about particular systems—for example, how forests will respond to climate change. Daniel Botkin, now at George Mason University, pioneered such work with early models of forests in which the measured growth characteristics of individual tree species provided the foundation. Subsequent research, by Botkin and by other researchers, has extended the range of these detailed forest growth *simulators* so that they can realistically represent the dynamics of a variety of specific forests. The simplest models, however, remain the vehicles of choice if we are interested in patterns that are common to a wide range of forests. Indeed, even when prediction for a specific forest is the goal, simplification of the detailed simulators is essential for making robust projections.[8]

A simple model gives us not only something we can understand easily but something that is likely to be robust, not sensitive to irrelevant detail. The philosophy of developing simple models has borne sweet fruit elsewhere in science, especially in physics. Some dismiss the importation of such approaches into biology as "physics envy," but this caviling misses the point. The power of simple models is that they isolate the essential truths, stripping away details that obscure the relevant dynamic features. There is beauty in the diversity of nature; indeed, that is the subject of this book. But there is elegance and basic understanding to be found in the general patterns that emerge from consideration of that diversity, and it is the illumination of those patterns that turns anecdotal reporting into science.[9]

It is the relevance of such models to a wide range of areas, from ecology and evolution to immunology to physics, that suggests that their predictions represent something basic. One must be careful, of course; removing too many details for the sake of exposing fundamental truths may lead to sterile and seemingly profound statements about nothing. Abstraction is like medication: if used sparingly and with care, it can do a great deal of good; overused and abused, it can be fatal.

The contact process is the model of immediate interest, so it is worth exploring its general predictions and the implications for the generation and maintenance of biodiversity. Clearly, if birth rates are much larger than death rates, sites (cells) do not stay empty for more than a negligible period of time; every site therefore is almost always occupied by one type or another. In other words, for very large birth rates, the contact process is no different from the voter model. As the birth rate is reduced, however, the degree of occupancy diminishes; eventually, as the birth rate for a particular species is made small enough, a threshold is reached below which extinction becomes virtually certain. Perhaps this is not surprising: it is clear that the birth rate must be bigger than the death rate for the population to survive. What is less obvious, however, is that it must be considerably larger than the death rate—as much as 60 percent larger for very big landscapes. This is a deep result proved by mathematicians exploring a subject called *percolation* theory and related topics in the study of cellular automata.

The birth rate must be much larger than the death rate, instead of simply larger, because occupied sites tend to be clustered together—a not surprising result of colonization being a local process. We are familiar with this local clustering in the spread of human diseases, which begin in individual households, spread to neighbors and schoolmates, and only later reach the general population. We observe it in the spread of mold on cheese left too long, allowing us to cut away the bad parts rather than give up on the whole hunk. And we see it in the spread of populations of plants and animals, like kudzu and zebra mussels and the Africanized bee. These start from a few introductions, spread to neighbors, and form clusters that become foci for expansion. Because of this clustering in the contact process, sites near occupied ones are more likely also to be occupied. Hence, many putative births will be wasted by occurring at already occupied sites. The implication is that more births are needed to compensate for those that are wasted, raising the bar for persistence.

Figure 5.3 illustrates this phenomenon with a snapshot of a landscape populated by three different species, after several hundred generations of the contact process. Note the high degree of clustering of individual species, and imagine the consequences for the potential spread of types. Potential parents in the middle of clusters cannot find any settlement sites for their offspring that are not already occupied by their own types. The other side of the coin is that this reduction in the rates at which types can invade tends to preserve inferior competitors longer, allowing them to find colonization sites themselves before they are eliminated locally. Overall, such clustering and spatial segregation thus helps to maintain biodiversity; indeed, it is probably the key contributor to sustaining biodiversity in most forests.

Local Diversity and Species-Area Curves

With these tools in hand, let's reconsider the species-area problem introduced earlier. My Princeton colleague Stephen Hubbell, an expert on the diversity of tropical forests, has amassed large data sets from a wide variety of different forests and revisited the fundamental problem of why regularities arise in the consideration of *species-area curves.* To explain these, he has proposed a simple model, which is just a minor variant of the contact process; again, the justification for invoking a simple model is in the conviction that when regularities

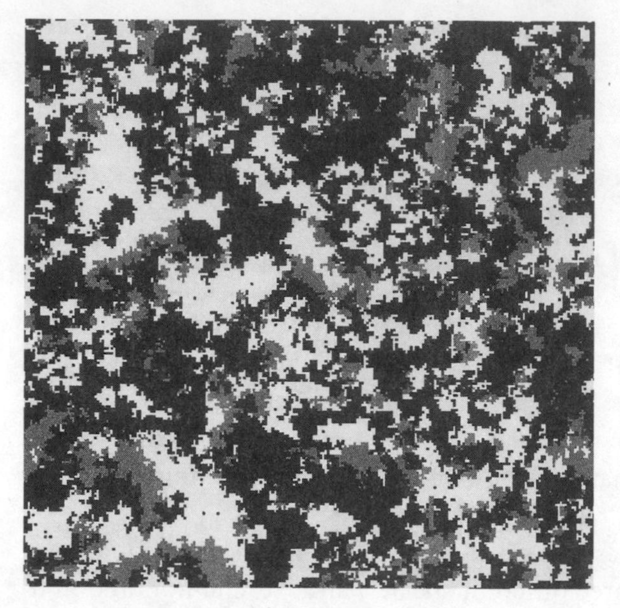

FIGURE 5.3
A snapshot of the contact process, simulating local competition involving three species on a broad landscape. Shadings denote different species. Simulation by Linda Buttel.

are so widespread, the reason must transcend the particulars of individual situations.

Hubbell imagines a landscape made up of various sites, each of which can support a number of plants of different species; thus, in his model, each site is large enough to support some degree of biodiversity on its own. The basics of the model mirror the contact process: individuals may die, or they may send offspring to colonize neighboring sites.

Hubbell adds a critical third ingredient: individuals of one type may spontaneously be replaced by individuals of a different type. This feature can be thought of as imitating either emigration from some unseen mainland source (on ecological time scales) or mutation (on evolutionary time scales). By varying the rate of spontaneous change, Hubbell is able to investigate the consequences of different degrees of isolation, or of different mutation rates.

Hubbell's model is beautifully simple, but difficult to analyze. Therefore, Rick Durrett and I, with the indefatigable assistance of Linda Buttel on the computer keys, have considered a slight variant that is more open to mathematical examination. Our model, which is termed the contact process with mutation, simply assumes a finer grid by breaking up Steve Hubbell's sites further into fundamental units, each the size of an individual plant. Thus, as in the basic multitype contact process, a site is either empty or occupied by a single individual. All other assumptions are the same.

From this simple model, in which all interactions are local, characteristic species-area relationships emerge. The larger the area considered, the larger the number of species found. Although many of the rates that drive the model are difficult to measure, it produces patterns that are in general quite consistent with observations. There are some discrepancies, pointing the way to the need for more elaborate models. The purpose of a simple model like this, however, is to see how much of the pattern can be explained without masses of assumptions, with the recognition that refinements are needed if one is to go the next step. What is remarkable is that the general shapes of these ubiquitous relationships can be explained with a model that assumes so little and therefore has very wide relevance. It does not mean that more is not going on in any particular system, but simply that parsimony argues against complicated explanations when simple ones are available.

The Assembly of Ecological Systems

Hubbell's model provides a way of studying at least some aspects of how ecological communities are put together at different space and time scales. To carry our story further, however, we must explore how phenomena at different scales relate to one another, and in particular the relationship between ecological and evolutionary processes.

Ecological systems are self-organizing—complex adaptive systems assembled from sets of available components as one would assemble a new computer system. Over long periods of time, feedback from computer users leads to changes in the lists of parts that are made available; in the same way, over long periods of time, feedback from ecological interactions in individual communities leads to changes in the lists or at least the availability of candidate species for ecosystem assembly. This involves evolutionary change, a topic I reserve for discussion in later chapters.

The self-assembly process is not uniquely specified, and individual communities take many twists and turns along the way. Indeed, even

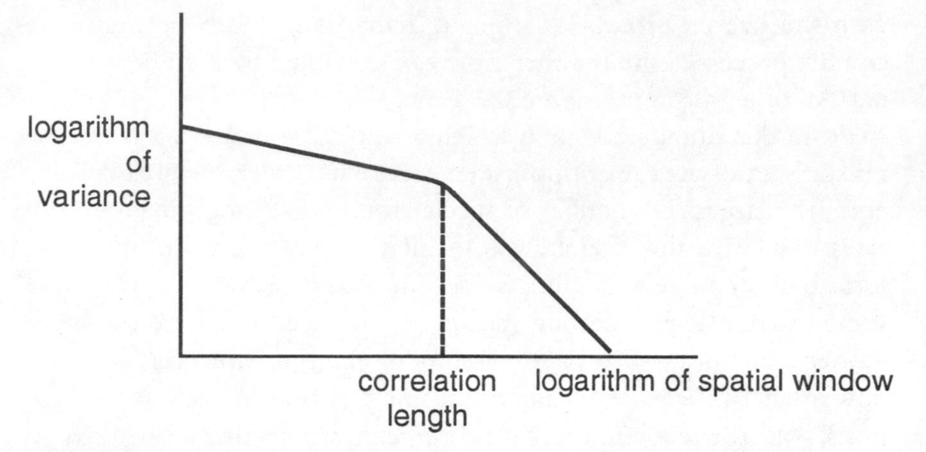

FIGURE 5.4 Theoretical Relationship Showing That Variability Decays with Spatial Perspective and That Predictability Increases. Illustration by Amy Bordvik.

once they are fully formed, ecological communities show a high degree of turnover and an impressive level of change. That was demonstrated most evidently in the classic studies in island biogeography discussed earlier, but any site in a forest would similarly exhibit high turnover in the occupants of the forest floor over long periods of time. In an equilibrium forest community, the climax community, variation in the kinds of trees that are present is constrained; many of the choices that were available in the early development of the forest are no longer open. Still, the variation that remains makes it impossible to forecast with certainty the local composition at fine scales; for purposes of prediction, one must change the scale, for example, by studying larger areas of the landscape.

A decade ago, Linda Buttel and I began to explore the consequences of such change in perceptual scale, by constructing a model of landscape development that shared much with Hubbell's model. The principal difference was that our model recognized a competitive hierarchy among species; we had in mind the various stages of successional development of the forest. We knew that the picture we got would be dependent on the window through which we peered—that is, on the size of the spatial area we examined. In particular, we knew that variability would go down, and hence predictability would go up, as we increased the size of the window. We were not prepared, however, for the clarity of the patterns that emerged. When we plotted the logarithm of variability against the logarithm of the size of the window, we found a remarkably linear relationship over a wide range of scales; variability was related to the size of the area considered according to a power law, just like the relationship between number of species and area described earlier. Here, however, the relationship was reversed—the larger the scale, the less the variability.

The exponents of the power laws become the slopes of the lines in Figure 5.4. Note that those slopes are negative, reflecting the reduced variability and increased predictability with larger scales. The exponents also are dependent on how tightly correlated, or interrelated, nearby sites are. Correlations are statistical measures of the degree to which different locations exhibit ups and downs at the same time. A zero correlation means that sites are independent; a positive correlation indicates that they tend to have their peaks and their valleys at the same times as one another, and a negative correlation, naturally, means just the reverse. The observed relationship holds only up to a distance called the *correlation length*, beyond which distance events are uncorrelated. This explains the elbow in Figure 5.4.

Long-term data sets of fluctuations in species abundance are hard to come by, making testing of this model difficult. However, lots of data exist for static distributions of species abundances and other ecosystem properties. For many dynamic processes, space and time are largely interchangeable: the variation that one can observe across space at any one time will eventually be seen at any site if we wait long enough. It makes sense therefore to look at the data again through different perceptual windows, but examining variation across space rather than across time.[10] Power laws turn out to be common; that is, the variation decreases as a negative power of window size. Eventually, for large enough window sizes, this relationship breaks down because sites within the window become uncorrelated.

This behavior is very reminiscent of critical phenomena—the magnetization of materials, or the conversion of water to ice—near phase transitions in physical systems, where such power laws are observed up to a certain scale. Most materials near such phase transitions exhibit power law scaling because events become correlated over wide areas. In *Cat's Cradle,* Kurt Vonnegut builds on this theme. Dr. Asa Breed, the somewhat mad scientist, explains the properties of a hypothetical crystalline form of water, ice-nine, which eventually destroys the world (delete this line from your memory if you have not yet read that wonderful book). From a tiny grain of ice-nine, Dr. Breed explains, copies of it form and stack, reproducing the pattern over and over again in a growing crystal.

Ice-nine never stopped spreading, frosting the feet of the sage Bokonon at book's end. More typically, in natural systems, the influence of an event at one point diminishes at distances away from that point, eventually disappearing altogether beyond the correlation length of the system. Trying to understand why ecosystems should be operating within this range—where scaling is important, suggestive of self-organized criticality—is what began my collaboration with Rick Durrett a decade ago, and it has kept us busy since. Together, we have learned a great deal, though much remains unanswered.

Ecosystem Organization

Despite the chance nature of local events, we have seen that regularities do appear in the organization of ecological systems. Patterns emerge from the interplay of local interactions and interchange among sites. In its broad features, this mirrors the prototype of pattern formation in biology, *embryogenesis,* the development of an organism from an embryo. But there is an important difference. In the development of an organism, selection is operating on the final product, shaping developmental pathways to reproduce a basic body plan, with no room for mistakes. Errors in this process result in birth defects and spontaneous abortions.

In contrast, there are many quite viable organizational plans for ecological systems; healthy forests, for example, can and do show great variation from one to the other in terms of species composition. Selection has operated at the level of the individual, not at the level of the whole forest. There surely are regularities, but these are typical of what one sees as emergent in any complex adaptive system. Large-scale patterns primarily *result from,* rather than drive, evolution at lower levels. Social and economic systems are similar in that their structures and macroscopic dynamics largely emerge from the selfish behaviors of individual agents rather than from top-down control. This is why the *Tragedy of the Commons* is such a ubiquitous problem; it is also one reason we have wars.

Regularities in ecological system structure and dynamics often come into focus when one takes a wider spatial perspective, as in moving from a clump of trees to a forest, or when one takes a longer temporal perspective, as in moving from weather to climate. Regularities may also appear as a consequence of a broadened organizational perspective, as in moving from individual species to functional groups, or simply to the total biodiversity of an area. For example, if one looks past simple species counts to the distribution of biodiversity, one finds canonical distributional patterns that seem independent of the system considered. The statistician Frank Preston discovered these basic patterns a half-century ago, and the distribution that bears his name remains one of the most accepted rules in ecology. Preston showed that abundances of individual species follow a *lognormal* distribution: that is, the logarithms of species abundances have a bell-shaped frequency distribution, much as would be expected for the exam grades in a large class. This is illustrated in Figure 5.5; as can be seen, there are relatively few rare species, and relatively few abundant ones. Most species have abundances in the middle of the range. Numerous investigators, from Preston to Robert May to Steve Hubbell, have developed mathematics to explain this regularity and to understand what determines the differences one does see (such as the width of the *bell curve*) in moving from one habitat to another. Closely related empirical studies, mentioned in the last chapter, examine the convergent evolution of communities that share similar physical environments, for example, arid ecosystems,[11] sorting out the effects of environment and history.

Simply listing species and their abundances is like describing a machine by its parts: it gives little insight into how the system works. How a machine functions is determined by how the pieces fit together, and how they interrelate in behavior; the same is true for an ecosystem. The wiring diagram for an ecosystem is embodied in its trophic web or food web, as introduced in the previous chapter. Just as Preston found statistical similarities among the distributions of abundances in a variety of ecosystems, others have examined and argued about the regularities that may be discovered in the topologies of webs across ecosystems.[12] How do those regularities arise? What accounts for the differences between ecosystems? How much of the variety is due to environmental variation, for example, physically controlled patterns of productivity, and how much represents the diver-

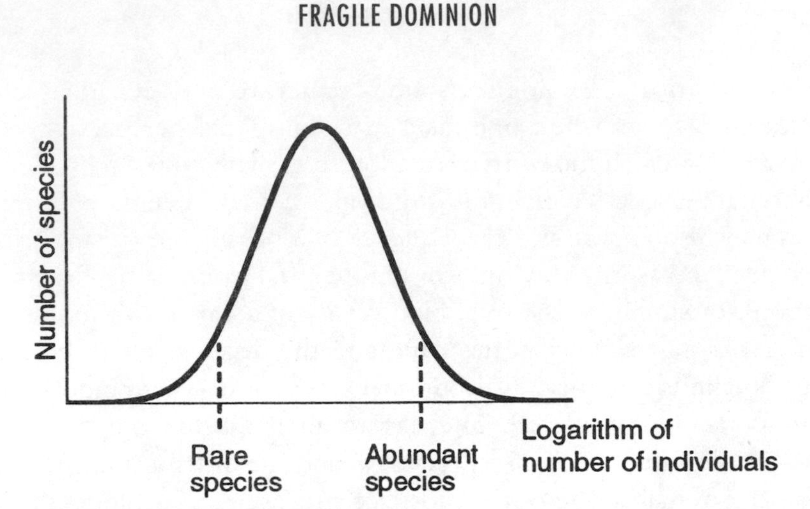

FIGURE 5.5 The Bell-shaped Lognormal Curve of Species Distribution. Illustration by Lee Worden.

sity that can emerge from the self-organization of a complex adaptive system? We cannot answer such questions just by statistical analysis of patterns; we must try to simulate and emulate the process of community assembly itself. Again, both controlled experiments and dynamical models can play a central role and have been used extensively to explore the consequences of various ideas about how food webs take form.[13]

David Tilman and his colleagues, in an important series of recent investigations, have shown that the productivity of ecosystems—the sum of all materials produced by plants—is to a large extent buffered against changes in species composition; that is, it is much less variable than are individual species abundances.[14] The reasons may be debated, but surely much of the constancy is explained by the fact that species simply replace each other in function. Similar studies have shown that marine fisheries have remained at reliable levels, although important individual stocks have come and gone.[15] Although total yields are at all-time highs, many fisheries favored by humans are severely threatened.[16] These data again illustrate the consequences of averaging over sectors, and the resultant smoothing of fluctuations. Broadening the perspective, whether over space, over time, or over organizational complexity, can lead to increased regularity and predictability.

Models and Experiments: Exploring Dynamics

It is reassuring to be able to reproduce broad general patterns from very simple models; such efforts can, however, take one just so far. The ultimate test of understanding is to be able to build a model that incorporates detailed knowledge of a particular system and to test its predictions against observations from that system. In a wide class of applications, where large numbers of individual agents are interacting with one another, that challenge means deriving testable predictions about the collective behavior of large ensembles of units from detailed information about how those units interact with one another. The classical success story in the physical sciences is the development of statistical mechanical theories to explain the thermodynamic behavior of fluids in terms of atomic physics. Other disciplines have faced similar challenges, though the rules of interaction are unfortunately not usually so well understood as are the principles of atomic physics. Indeed, this problem of explaining dynamics at the macro level in terms of interactions at the micro level is the crux of the matter in the study of any complex adaptive system, from socioeconomics to the brain. In ecological systems, the issue arises repeatedly; it is the bridge between molecular biology and organismal biology, between organismal biology and population biology, between population biology and ecosystem science.

Statistical mechanical theories have long intrigued theoreticians in ecology. Extremely clever work by researchers such as Ed Kerner and Egbert Leigh in the 1950s and 1960s used constructs from physics, in particular the Gibbs ensemble, to develop a formal theory of ecological organization; indeed, many aspects of their framework could be traced back to the master, Vito Volterra.[17] The elegance of their work stimulated a brief flurry of research by others; the assumptions, however, clearly were violated in fundamental ways in ecological systems, and the approach proved to be a blind alley.

Robert MacArthur, much intrigued by such efforts, began to develop still fairly general models of ecological organization, starting from realistic assumptions about the dynamics of individual species and their interactions with one another, and attempting to derive general principles about limits to how many species could be packed into a community, and about the efficiency of resource use. His lead paper in the inaugural issue of the now venerable journal *Theoretical Population Biology* was a marvel of innovation and creation. Such modeling was suggestive and stimulating, but even in its best form it was never intended to be testable in any particular system.

While MacArthur was developing his theories, a quite different philosophy was driving other theoretical research on the structure and functioning of ecosystems. MacArthur was a population biologist—an evolutionary ecologist who approached ecosystems from the bottom up. His work was inexpensive and modest in scale, and he tended to work alone or with a few selected collaborators. MacArthur was committed to the notion that ecosystem properties would emerge from consideration of the dynamics of populations and would be illuminated by an evolutionary perspective derived from studying populations. At about the same time, the International Biological Program (IBP) brought together large numbers of researchers to develop massive computer models. While MacArthur and others in the same tradition dealt in generalities and universal principles, the IBP dealt with specific ecosystems. The legacy of the IBP was mixed, reflecting the fact that it involved large networks of teams of researchers. Much basic empirical work was carried out that provided a database far beyond what had previously been available. Furthermore, some of the deepest thinkers in the IBP, such as Robert O'Neill at Oak Ridge National Laboratory, used the modeling efforts as a springboard to understanding fundamental principles of ecology as well as fundamental issues in modeling. The driving force behind the IBP, however, was the creation of large, highly detailed models that could take advantage of what then seemed to be great computing power. Though beginning from a top-down perspective, in which ecosystem processes were the objects of interest, many IBP models nonetheless foundered by accepting the notion that the more detail that was included in the models, the better they would represent reality. The IBP models provided counterpoint to MacArthur's efforts in that they were tailored, in painful detail, to particular systems. Models of such overwhelming complexity, however, are so demanding of data—sensitive to small differences in the measurement of particular variables and capable of expanding on and propagating their own errors—that they have little value beyond providing a framework on which to hang information that has been gathered in support of them. Daniel Janzen, a prominent population biologist, complained that these models got "all the nouns, but none of the verbs."[18]

As a result of these differing philosophies about how to approach the question of ecosystem organization, the gap between population biology and ecosystem science remained too wide for too long. Indeed, it was not theoretical work but stunning empirical successes that began to drive a merger of the best of both disciplines. Evelyn Hutchinson saw no sharp boundaries in his work on lakes, and the influence of his approach could be seen clearly in the best work of population biologists and ecosystem scientists alike. Lake Washington became a model for how to restore a polluted ecosystem.[19] The Hubbard Brook Experimental Forest became the most important experimental ecological facility in the world, providing understanding of the ecosystem consequences of clear-cutting forests and the first evidence of acid precipitation in North America.[20] Robert Paine's landmark studies in the rocky intertidal provided a framework for understanding the importance of disturbance in structuring communities; similar insights emerged from Hubbard Brook. These and other studies showed that there were fundamental linkages between levels of organization. Ecosystem properties were shown to emerge from the dynamics and interactions of populations; in turn, populations could be understood only within the context of the ecosystems in which they occurred. There were profound implications for modeling as well. The dynamics of systems, and in particular their capacity to recover from perturbations, could not be understood from observations alone; experimental manipulations, controlled or otherwise, were essential to exposing the nonlinearities that governed system resilience.

The work of a few pioneers (Bormann and Likens, Paine, Schindler) stimulated experimental approaches to ecosystem studies that began to produce the data sets that could provide the critical tests for models. Botkin's forest growth simulator, *JABOWA,* spawned healthy offspring,[21] which are today proving valuable in explorations of the environmental effects of greenhouse gases. Models such as these are imitators of landscape development, tailored to particular forests, such as those of the northeastern United States.[22] In structure, they are applicable to a broader class of forests, even tropical forests and chaparral, with little modification. In effect, they are flexible because they are elaborate versions of much simpler interacting particle models, such as the contact process. Sites on a landscape are occupied by trees of particular types, which grow and cast shade over their neighbors or otherwise compete for resources. Such competitive interactions affect the future growth of trees, as well as their reproduction, mortality, and dispersal. The basis for such models is a set of empirical relationships for those fundamental processes, derived and measured to apply to particular forests.

When one of the latest of these models, *SORTIE*, the creation of my colleague Stephen Pacala and his collaborators Charles Canham and John Silander, is allowed to run on the computer, patterns of clustering of trees begin to emerge, much as for the multitype contact process. The result, as can be seen from the snapshot in Figure 5.6, is something that looks like a scale model of a forest. The landscape is a mosaic of stylized trees, huddled together in clumps of their own type. Over time, the picture changes; the clumps are more or less the same, but they move around. It is, indeed, as realistic an imitation of a forest as one can imagine. In its explicit predictions, however, it offers more information than we should expect to be able to know about a system. The future exact locations of trees of particular species are in reality unknowable; what is knowable are averages of abundances over broad spatial areas, or long time scales, and perhaps similar information about the degree of clumping. The challenge in describing any system of this sort is to separate the unknowable from what is simply unknown but ultimately knowable.[23]

SORTIE is solidly grounded in empirical detail; as a tree grows, the model assesses the light hitting it from each of 256 directions. Such precision may be important for an individual tree but surely cannot matter at the level of a landscape; if it did, all hope of prediction would be lost. Imagine the dilemma if your physician needed to be able to predict how every cell in your body would react to a drug, and how those individual responses would translate into an organismal response. If individual cells were independent, that might be possible, at least in the sense of statistical averages. But the body is a complex adaptive system and thus governed by nonlinearities that resist predictive sorties. In the same way, ecosystems are complex adaptive systems; if all the fine-scale detail that is built into predictive SORTIEs were essential, the model would be useless. It is tempting

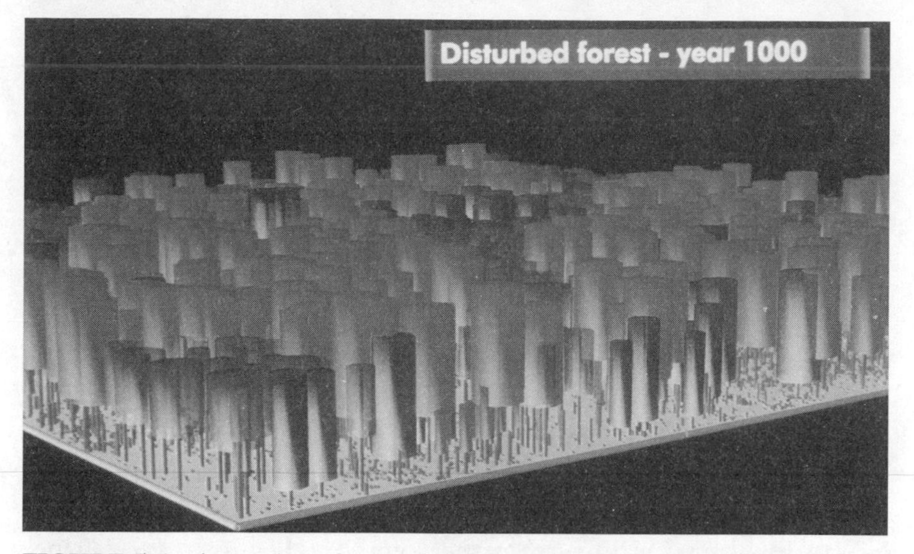

FIGURE 5.6 A snapshot of a simulation model of the dynamics of the Great Mountain Forest (Connecticut). Simulation shows forest at year 1000, in the presence of local disturbance due to treefall. Model developed by Pacala, Silander, and Canham, "Forest Models Defined by Field Measurements"; simulation by Douglas Deutschman and Linda Buttel on the Cornell University supercomputer.

to believe that the surest way to a better model is to get more and more accurate models of the pieces, and that if the pieces are right, all else will follow. Not true. Indeed, the pieces must contain some of the truth, but the surest way to a better model is to string together the pieces so that small errors in the fine-scale details don't matter much.

Designing predictive models so that small errors in detail will be forgiven is not the province of mathematicians and computer scientists alone. This is, after all, how we run our lives. To survive, one must be constantly making predictions and acting on them. Who is this person approaching me? Is it a friend or a foe? Should I continue to approach or take evasive action? A wealth of detail is available, too much, indeed, for me to process completely. My response system is a complex adaptive system, shaped in part by evolution but honed by experience. Its first job is to abstract, to categorize information by suppressing unnecessary detail[24]; the way it does that is continually being updated, based on the success or failure of previous decisions. The most successful schemes—indeed, the only successful schemes—

are ones that suppress irrelevant detail to allow decisions to be made quickly and robustly. The same principles need to be applied to computer models: one must abstract, condense, and simplify in order to suppress irrelevant detail and allow the message to surface from a sea of noise. There are various ways to do this, from making mathematical analyses to simply running the model over and over again, providing a library of experiences similar to what we use in honing our responses to environmental stimuli. It is in this process of abstraction and simplification that a science of prediction emerges.

Forest models provide beautiful examples of how one can use simulation techniques to understand how ecological systems self-organize. They are grounded in measurements and testable against historical records of particular forests. Eventually a characteristic forest, to some extent independent of what the theoretical forest looks like at the beginning of time in the computer, begins to take shape. Driven by local patterns of dispersal and interspecific competition, the landscape becomes organized into clusters of trees of particular types. The existence of those clusters changes the dynamic, since now species interact more with their own type than with other types. The world as seen by an individual tree changes; this is manifest as nonlinearity in the landscape dynamic and helps shape the future growth of the forest. Depending on the details, the entire development of the forest may be shaped by chance events early in succession, but many models are sufficiently constrained in their scope that this is not observed. In any case, it is the processes of local disturbance and renewal that maintain the diversity that we see in systems, and in most models of those systems. Local variability and heterogeneity provide the material for change by which the system adapts and maintains its resilience. This point is at the center of the theory of complex adaptive systems, and one to which I shall return repeatedly. It is key to the writings of a diversity of scientists who have influenced my own thoughts about adaptation, including especially Larry Slobodkin, Crawford "Buzz" Holling, and John Holland. Indeed, the subtitle of Holland's book in this series, *How Adaptation Builds Complexity,* captures this theme elegantly.[25]

Forest models are not alone in demonstrating such general features. Grasslands and the rocky intertidal are much like forests and other complex adaptive systems in the broad features of self-organization;

local disturbance leads to clustering and nonlinearity, and to the maintenance of biodiversity.

In the rocky intertidal, my own work years ago with Bob Paine focused on a relatively small area of the northwest coast, small enough that many of the key factors were extrinsic to the area studied.[26] Disturbance, largely due to wave action, played a role similar to what Alexander S. Watt elegantly described for forest systems, ripping holes in mussel beds and renewing a limiting resource.[27] Diversity is maintained because of the global certainty of local uncertainty; that is, the continual renewal of opportunity through localized and random disturbances creates evolutionary opportunity for locally transient species.

Paine and I found that the logarithms of sizes of disturbances had a frequency distribution described by a bell curve. This is what is to be expected, according to mathematical theory, if a large number of independent extrinsic factors are contributing to the formation of patches; in contrast, were disturbances the manifestation of criticality, the evidence of a system generating its own catastrophes, a power law distribution should be expected, as in Per Bak's theory. Once these local extrinsic disturbances occur, however, self-organization would still take over, shaping the community's development. The areas we studied are among those on the outer coast in which extrinsic disturbance is most severe; other areas, such as in the protected waters of the San Juan Islands, or off the California coast where Joan Roughgarden has focused her work, may demonstrate patterns of disturbance that are more the result of self-organized criticality. This is indeed suggested by the biology: buffered from the effects of strong wave forces, the invertebrates in more protected areas are likely to live longer, to become more crowded, and to become hummocked or stacked on top of one another. It would be fascinating to compare the size distributions in such areas with those of the outer coast to see whether power law distributions describe the data better than the lognormal, perhaps suggesting evidence for self-organized criticality. At this point, such a suggestion is no more than idle speculation.

Similar phenomena are also evident in other complex adaptive systems. Joshua Epstein and Robert Axtell have explored the dynamics of social agents on resource landscapes that they call "sugarscapes"—sugar is, in their model, the sought-after good—demonstrating that many of the regularities that emerge in the distribution of wealth and resources can be understood as the necessary consequences of the self-organization of human societies, special forms of ecological systems. Indeed, Epstein and Axtell explicitly treat economic systems as ecological systems, through simulations of the behaviors of individual agents finding and exploiting resources. Not surprisingly, the same sorts of patterns of aggregation and self-organization that are evident in simulations of ecosystem dynamics reappear in Josh and Rob's work.

Brian Arthur has long emphasized the importance of nonlinearities and path dependence in the development of organization in economic systems, and Steven Durlauf focused much of the economic program at the Santa Fe Institute on the problem of developing an appropriate statistical mechanics of economic systems. As always, for all of these investigations, a fundamental challenge is to separate extrinsic and intrinsic influences on the development of pattern.

The Ecological Theater and the Evolutionary Play

Few scientific disciplines have advanced as rapidly as ecology. Thanks to research carried out by a variety of scientists, a newly minted ecologist today understands a great deal more about the dynamics and organization of ecosystems than most experts did a half-century ago. At small spatial scales, for example, we know that there is a great deal of uncertainty and variability in species composition, but that that uncertainty largely disappears as attention moves to landscapes. Indeed, local disturbance and uncertainty—such as is provided, for example, by small fires—is the main force renewing diversity and maintaining species in the system that would be outcompeted in less dynamic landscapes. Understanding this has changed the way we manage fires on public lands. More generally, the theory of island biogeography, originally developed for mangrove keys in the ocean, has proved a powerful tool for understanding how biodiversity is maintained in forests, mussel beds, grasslands—and, indeed, in virtually any ecosystem.

The most important challenge for ecologists remains to understand the linkages between what is going on at the level of the physiology and behavior of individual organisms and emergent properties such as the productivity and resiliency of ecosystems.

Ecosystems are prototypical examples of complex adaptive systems, with strong similarities, for example, to socioeconomic systems. Through modeling exercises informed by experiment, we have seen that we can learn much about dynamics, and explain a great deal about statics, by building ecological systems up from their components and developing an equivalent statistical mechanics. Such approaches have transformed our understanding.

Self-organization on ecological time scales, however, which has been the focus of this chapter, is no different in kind from self-organization over evolutionary time; indeed, the attempt to distinguish these sharply is only a pedagogical simplification and is confounded by the existence of a continuum of space, time, and organizational scales inextricably tied up with one another. If ecological communities are assembled from components shaped by evolution, how does the sum total of those ecological experiences feed back to change the evolutionary context? Hutchinson, to whom ecological theory owes so much, set the problem in his wonderful book *The Ecological Theater and the Evolutionary Play*, from which this section of my chapter derives its title. Inspired by that metaphor, I turn in the next chapter to try to put these thoughts within their evolutionary framework.

6

THE EVOLUTION OF BIODIVERSITY

Playing with Legos is for beginners. The Lego metaphor given earlier imagined a single player assembling an ecological community. The truth of the matter is that community assembly is really an interaction among multiple players; for the advanced class, now, we must move to a competitive game situation, a far better analogy.

The assembly of ecological communities is more like a giant Scrabble match. For both situations, the game consists of drawing tiles from a master set and placing them on a board, a landscape, in which only local rules of interaction constrain a remarkable diversity of possible realizations of the game. The game of Scrabble requires players to take turns, constructing a crossword puzzle in which every word formed is part of the standard language. Though there technically are a finite number of ways that pieces can be placed on the board, the actual number is so high that every game may be regarded as a unique event. One cannot hope to predict at the beginning the exact placement of letters, but one can guess rather confidently that there will be nearly one hundred tiles on the board at the end of the game, and not be off by much. In the same way, ecological communities assemble themselves in unpredictable ways from master sets of species existing in some larger context. The exact composition of such a community is impossible to predict, but macroscopic descriptors such as the equilibrium number of species (tiles) can be estimated fairly well, even for areas that have never been studied before. Such estimates could be based on comparisons with similar habitats and the use of graphs that relate the number of species to the size of an area. There is one important difference worth mentioning, however, since it could lead to a wildly successful improvement of Scrabble. My creation, Ecological Scrabble, is a game that never ends. The rules permit not only the placement of tiles on empty spaces but also the substitution of tiles for ones that are already on the board. This cannot be done randomly; the replacement of tiles must leave a configuration that makes sense in terms of the local interactions. This would make for a quite challenging extension of the game of Scrabble and adds new dimensions to the picture.

Who determines the set of species from which tiles are to be drawn? In the case of Scrabble, we know the answer: there are real people who made and maybe still make such decisions, based on the frequency of letters in the words in whatever language a version of Scrabble is being designed. In principle, initial guesses at what would make for good games of Scrabble could be changed after experience with many games showed that the preference for letter use did not mirror that found in a dictionary. The rules of Scrabble assign more points for the use of a rare letter, so these are played preferentially in key situations. It is not at all clear what available letter frequencies would make for the most exciting game; those frequencies, and the values given to letters, could be changed over time to reflect the experiences people have had playing with particular letter sets. I have no idea how this could be done or was done by Selchow and Righter, which marketed Scrabble, but I am sure that something like it must have happened.

Imagine how this would have worked if Scrabble had been designed the way Ecological Scrabble was. Picture a big factory with hundreds of workers playing versions of Scrabble with different-size boards and different frequencies and valuations of tiles; each player has an attached happiness meter that indicates satisfaction with the game. Over time, the sets that generate the most happiness are distributed to more and more people until everyone has the same version, which on the whole generates the most happiness. That version is then marketed. Indeed, there is some advantage to marketing a number of different versions to reflect the fact that people differ in their preferences. To some extent this is done with regular Scrabble—

there are versions in different languages and kids' versions—but the realities of marketing and a desire for standards constrain the real diversity of choices.

The point of this somewhat contrived example is a fundamental one. Even the design of a product, which we tend to think of as specified by an engineer, is to some extent part of a complex adaptive system, which learns from experiments, typically carried out at scales far below the level of the total system. In the design of a board game, an initial design stage is followed by an improvement stage that consists of the playing out of many versions of the game. Improvement depends critically on heterogeneity and diversity—that is, on the existence of many alternatives to choose from. Learning is efficient because one can compare the success of one version of the game with that of another. The system as a whole changes over time because of a process of selection, essentially natural selection, among its parts.

In game design there are people who actually make the decisions about how to use the information filtering back about customer satisfaction, but in theory this process could be automated. A training program for a machine learning to play a game, for example, could change the probability of certain situational responses based on measures of overall success associated with that response; in the same way, we adjust our own responses to various situations based on experiences with similar situations in the past. A common feature of all of these examples is that the measure of success is integrated over a number of individual decisions or responses that have been made, and there is a statistical sorting that needs to be done.

The assembly of an ecological community is in essence part of such an automated design process that is continually changing the available components and their roles. The experiences of a type over all communities affect its total reproductive success. Those types that leave the most offspring come to be represented more in future generations—this is how the set of available types is determined. This is evolution writ large, or at least writ larger than what might go on within an individual community. In this way, the set of available colonists constrains community assembly, and in turn, the collective experiences of colonists across all communities refine and reshape the colonist pool. The process is hierarchical: a *meta-community*—that is, an ensemble of communities—changes in composition on a slow time scale as the result of interactions within many individual communities on fast time scales.

Islands and Evolution

The clearest examples of meta-communities are island chains, which are wonderful places to study evolution. The Galápagos Islands, for example, have provided a remarkable showcase for evolutionary studies since Charles Darwin and the *Beagle* landed there. Peter and Rosemary Grant, colleagues of mine at Princeton, are the leading contemporary Galápagos researchers, carrying out modern classic studies of evolution in Darwin's finches.[1] As I discuss in more detail later in this chapter, the Grants have shown conclusively that important evolutionary change can occur over remarkably short time scales. The island setting also makes clear the importance of chance and history in creating ecological and evolutionary differences among islands.

Any island chain, indeed, presents an ideal laboratory for studying the interplay of local ecological processes and global evolutionary change, because of the sharp division of the environment into discrete and distinct habitats. Although such features are perhaps clearest in oceanic archipelagoes, the meta-community structure is a fairly typical one even in terrestrial habitats. Trees exist in patches of forest habitat, adrift in a sea of urban development and farms; agricultural fields and lakes similarly constitute their own archipelagoes. Indeed, it is hard to think of many examples for which the island model is not appropriate for the examination of the evolution of indigenous biotas.

The dynamic is played out on smaller scales as well; within an individual forest, islands are continually appearing and disappearing, creating temporary islands of opportunity. Over evolutionary time, species acquire attributes that specialize them to particular stages in the local successional process. Early successional species develop adaptations, such as high dispersal rates or dormancy, that allow them to find new habitats quickly. Late successional species, on the other hand, develop competitive traits, such as the ability to grow in the shade and overtop other species, that allow them to displace the early colonists. Such adaptations emerge from the collective realization of the unfolding of the successional dynamic over and over again at local scales. In this way, the forest biota is shaped; in the same way, agricultural fields acquire their characteristic pest species, lakes their resident fauna and flora, and animal populations their parasite burdens.

This hierarchical perspective, of course, provides the context, but not the mechanisms, for evolutionary change. How, indeed, does evolution happen? At the most primitive level, the story is simple in detail, overwhelming in its force and elegance. Evolution by natural selection can take place only within a population that is genetically variable for some trait that influences the fitness—that is, the reproductive success—of its bearer. As one generation is replaced by another, those forms of the gene that impart higher fitness increase in frequency in the population, displacing the less fit. That is the essence of the story. It becomes complicated in its specifics because genes do not act independently of one another, and because the effect that a gene has on an individual's fitness may change over time. The basics, however, remain constant: chance and choice.

The details of evolutionary change are encoded in the genes, the basic units of selection. Evolutionary systems, however, are the epitome of complex adaptive systems: change is manifest at a spectrum of hierarchical levels from the gene to the organism to the population and beyond. Understanding how evolution works, therefore, requires integrating molecular and population perspectives by elucidating the dynamics of populations of genes. This is the province of *population genetics,* one of the most venerable branches of biology.

The Genetic Basis of Evolutionary Change

In the first half of the twentieth century, three intellectual giants—Sewall Wright, Sir Ronald Fisher, and J.B.S. Haldane—collectively laid the foundations for the theory of population genetics. One of Wright's most enduring contributions illustrated the essentials of the evolutionary process through an evocative visual metaphor, the *adaptive landscape.* Wright first presented this idea in a major invited address at the landmark 1932 International Congress of Genetics, organized in my old hometown, Ithaca, New York.

My close friend William Provine's account of the congress makes for fascinating reading.[2] Wright, Fisher, and Haldane were all invited to lecture in a special session. The session was organized because everyone recognized that these three geniuses were revolutionizing the subject, but few could understand the details of their papers. The session was to provide a format for them to present their landmark research in ways that even a layperson could understand. The problem of communicating mathematical ideas to a general audience remains a daunting one today, and the fathers of population genetics were no less challenged.

Wright's effort at communication was to imagine a mountainous landscape, with a rugged topography characterized by huge numbers of peaks and valleys. A species was represented as moving on this landscape much as a ball would, but seeking peaks rather than following gravity to the wells. Turning the landscape upside down would have made the metaphor of a ball on a landscape a natural one but would have lost the notion that the peaks represented adaptive peaks, points where fitness was higher than at any nearby neighboring peaks. Wright wrote:

> The central problem of evolution as I see it is that of a mechanism by which the species may continually find its way from lower to higher peaks in such a field. In order that this may occur, there must be some trial and error mechanism on a grand scale by which the species may explore the region surrounding the small portion of the field which it occupies.[3]

On the one hand, this simple conceptualization has been a remarkably useful pedagogical tool. On the other hand, it has some serious technical limitations, to which I return later.

Wright had strong roots in animal breeding, and the notion of the adaptive landscape is well tailored for the challenges that face a breeder. Under artificial selection, the breeder decides which types are to be preferred—that is, to be selected. In effect, fitnesses are assigned to specific types in the sense that only those that are favored are allowed to breed. An adaptive landscape is similarly defined: a flowing surface suspended above the various gene combinations available, as in Figure 6.1. If, for example, the breeder is selecting for coat color, the surface above gene combinations that produce the most favored coat colors represents the peaks of the surface. Through generations of selection, the population is driven to a local peak, though not necessarily to the highest of all peaks. The landscape itself remains constant until the breeder chooses to change it, selecting for other features of interest or responding to new challenges.

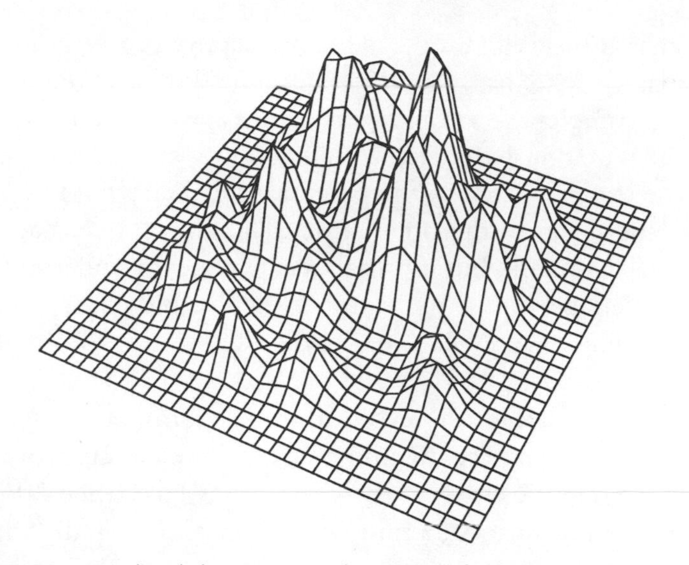

FIGURE 6.1 An Adaptive Landscape with Many Peaks and Valleys. Illustration by Lee Worden.

The metaphor of the adaptive landscape allows one to visualize a number of fundamental features of evolution in nature. As François Jacob emphasized, evolutionary change is highly contingent on historical accidents, which influence the multitude of choices to be made among the possible peaks to be reached.[4] The process of achieving a peak is one of gradual and steady hill climbing. Some peaks may be low and draw from very small regions; such narrow peaks are therefore rarely achieved, because few pathways lead to them. Furthermore, populations that are perched on low peaks are vulnerable to invasion: the infusion of novel types by immigration or mutation can easily destabilize such a fragile equilibrium, introducing *genotypes* (individual genetic types) that sit on the slopes of distant mountains, thereby catapulting the population toward a new equilibrium.

The adaptive landscape provides a rich metaphor. Before exposing its flaws, therefore, I turn in the next section to exploration of some of its implications for understanding the evolutionary process.

Evolutionary Flows on Landscapes

Stuart Kauffman and I, in 1987, became interested in deriving some of the further implications of the adaptive landscape for studying evolutionary change, as well as for looking at adaptive processes in other complex adaptive systems, such as the immune system.[5] We began by considering completely *uncorrelated* landscapes—that is, ones in which slight mutational changes would produce types whose fitness was no more closely related to the parental type than would be a type chosen at random from the population. This is not a very accurate model for real evolution, but it provides a baseline for understanding. In such simple landscapes, the number of possible peaks becomes very large, leading to an evolutionary process in which history plays a dominant role.

Consider, for example, the evolution of *proteins,* large chains of *amino acids.* Since there are 20 amino acids, there are 20 choices for each position along the chain. If the chain has N links, then 20^N possible proteins can be formed, a number that increases rapidly with N. Evolution typically operates on very high-dimensional spaces. The number of local peaks may also become astronomically large in realistic situations. If fitnesses are assigned randomly to *polypeptide chains* (chains of amino acids), and if $N = 50$ (that is, the chains are made up of 50 amino acids each), we showed that about one-tenth of 1 percent of all chains have fitnesses that are local maxima; that is, about one-tenth of 1 percent have fitnesses that are higher than any other chains that could be formed by changing exactly one amino acid. One-tenth of 1 percent may seem small, but it is a percentage of a very large number. Hence, the number of possible local peaks is astronomical even for short-chain molecules; proteins typically involve much longer chains and so exist in even larger configuration spaces. An evolutionary walk on an uncorrelated landscape of possible proteins thus has a dizzying range of possible pathways that it can follow; it would be hopeless to try to predict its eventual equilibrium.

There are many problems with this example, of course. First of all, while it is true that a type that is more fit than its one-step neighbors is technically associated with a local peak on the fitness surface, such a peak is more a pimple than a mountain unless that type exercises similar dominance over a larger neighborhood. Populations would not stay perched for long on pimples, since simultaneous mutations at two amino acid loci could allow evolution to explore much larger neighborhoods. Even more fundamentally, real landscapes are not uncorrelated; changing one amino acid in a chain is likely to produce a new protein with fitness not so different from that of the original type. The effect of this is to produce landscapes with far fewer peaks, though the number still may be quite large.

To explore how this hill-climbing process might work in practice for correlated and therefore more realistic landscapes, we asked how such an evolutionary process would deal with one of the classic problems in the theory of optimization, the *traveling salesman problem*.[6] Our objective was not to find the most effective algorithm for solving the problem; for that purpose far better techniques already were known, including pioneering work on *linear programming* by one of my own mentors, George Bernard Dantzig. Rather, we wanted to understand how efficient an evolutionary algorithm would be in dealing with it. We did not realize it originally, but *stochastic optimization* methods[7] very similar to those that we explored actually represent the state of the art for large ensembles of cities.[8]

The traveling salesman problem has attracted so much attention because it is a prototype of a wider class of problems of far-reaching importance in theory and practice. The formulation assumes that a salesman must visit each of a large set of cities on a sales tour, and it asks what order of visiting will result in the shortest amount of total travel. It is easy to see that for even a moderate number of cities the complexity of this problem is forbidding: the number of possible tours increases factorially with the number of cities. An efficient way to search the possible space of tours is essential. We idealized the various tours as distinct genetic types; that is, we thought of the arrangement of cities along a tour as being equivalent to the arrangement of genes along a chromosome. The fitness of a tour is determined by its length: for example, since shorter tours are more desirable, one might define fitness as the reciprocal of the length of a tour. In this way, a landscape was defined, consisting of the universe of possible tours and their associated fitnesses.

To have an evolutionary dynamic on this landscape, one needs a source of variation, analogous, for example, to mutation in biological systems. For us, mutation consisted of a variety of mechanisms for rearranging the order in which cities would be visited; the simplest involved switching the position of one city with that of another. So in each generation each tour begat another tour, selected as the fittest among candidates that included the original tour and various mutants. The new tour might resemble its progenitor, or it might differ in the order of a few cities if that rearrangement produced a shorter tour. The simplicity of this scheme mirrors the simplicity of the evolutionary process itself: generate variants (chance); select the best (choice). And it works. Beginning from randomly defined genotypes—that is, randomly ordered visits to the required cities—we found this evolutionary mimic to be capable of rapid and substantial improvement until a local optimum was achieved. The details depend on the rules of mutation and selection, but in general, the landscape had several peaks, leading to a strong historical contingency in the dynamic optimization process. That is, depending on how things got started, one could end up at a variety of different shorter tours that were resistant to further improvement.

Evolution as Optimization

The notion of optimization on landscapes as a metaphor for the evolutionary process has some fundamental weaknesses. The main problem is that environments change, so that fitnesses vary and landscapes shift. Trying to climb hills on such a landscape is like trying to stay on the crest of a wave.

Under some conditions, however, the model is not a bad one. If environmental change is slow, as it is in the world of the breeder, the notion of a fixed landscape makes sense, at least for short periods of time. Indeed, the strongest formal justification for Wright's adaptive landscape is to be found in the work of his rival, Sir Ronald Fisher. Fisher showed that, in the simplest cases, the average fitness of all individuals in a population inexorably increases over time until the population comes to rest at some local optimum—that is, until it sits on a peak in the adaptive landscape. Fisher's *fundamental theorem of natural selection* applies to populations rather than individuals, so it is a bit different from Wright's original formulation; but Wright's representation of his ideas went through a comparable evolution of its own, so Fisher's theorem is not far off the mark.[9]

The problem is not so easily solved, however. Unlike individuals, populations do not have a unique fitness associated with them; rather, they are made up of collections of individuals that differ in their fitnesses. One may attempt to characterize a population by the average of the fitnesses of the individuals that make it up, but that simply disguises the existence of variation. In reality, one should think of a population, not as a point on the adaptive landscape, but as a cloud of points distributed about a central position. If the center sits on some narrow peak, the population distribution reaches down the sides of the hill, and perhaps a bit up the next one. Consider, for example, the emergence of complex biological structures—the eyes of higher organisms, or adaptations for feeding or sexual union—that require mutational changes at several different loci. Such adaptations have often mystified biologists, because the intermediate steps between an existing type and an evolutionary breakthrough may all represent selectively inferior types. If mutation maintains variation, however, thereby preserving these inferior types against selective extinction, eventually successful combinations can be found that allow evolutionary jumps.

The various mechanisms, such as mutation and sexual recombination, that maintain variation within populations change the evolutionary dynamic fundamentally. Evolution proceeds by the reproductive success of individuals and by virtue of the variation within populations; suppressing variation eliminates the sine qua non of evolution. The model that Stu Kauffman and I first explored, by ignoring variation within populations, understated the ability of evolution to escape local peaks and find other optima. We knew about this shortcoming and attempted to address it by allowing for the occasional replacement of more fit types by less fit types. In this way, the optimization process could escape narrowly defined local optima. Our approach was phenomenological, one meant to imitate the consequences of a variety of mechanisms, such as mutation (spontaneous change at a single genetic locus) and recombination (change arising from novel combinations of genes at different loci), that preserve variation.[10]

Sewall Wright, in his landmark 1931 paper, presented a fundamental mechanism by which populations could escape from local peaks.[11] He argued that a large population will be broken up into smaller populations, distributed over a variety of habitats. The local adaptation of these subpopulations to differing conditions helps maintain a heterogeneity in the population that, through emigration and sexual recombination, generates a much larger range of variants than could mutation alone. Wright argued that this *shifting balance theory* provides one important mechanism by which variation is maintained and the search for optima is enhanced.

There are deeper problems with the notion of the adaptive landscape, however.[12] In nature there is no breeder to hold the fitness landscape constant. Environmental change is constantly shifting the background against which selection is taking place; thus, adaptation occurs on a landscape that is in a perpetual state of oscillation, with peaks dissolving into valleys and new peaks arising from former valleys. Worse, those fluctuations result not from *extrinsic* influences alone but also as the result of evolution itself. For example, as evolution shapes a population of plants to exploit a particular type of habitat, competition in that habitat type consequently increases. With increased competition comes decreased fitness, making other habitat types more desirable than they were previously. The fitness landscape has shifted, like a waterbed, from the weight of too many individuals perched on what once were peaks.

This phenomenon is known as *frequency dependence* when it operates within populations. Frequency dependence means simply that fitnesses are not constant but instead depend on the frequencies of types within a population. For example, on a particular island selection may initially favor birds with large bills that allow exploitation of seeds of a particular type. As the frequency of birds with large bills increases, the seed resource becomes overexploited and the fitness of the large-billed birds is diminished.

Frequency dependence can operate in many ways to decrease the fitness of common types. Competition for limited resources provides the most obvious example, but it is not unique. Since predators may preferentially search for common prey types, increasing frequency of a particular type also increases predation risk and lowers fitness. Similarly, disease risk may increase with the frequency of a type, since the probability of transmission becomes higher. On the other hand, frequency dependence can also operate to enhance the fitness of common types. It is quite common in many species, for example, that individuals prefer to mate with their own kind. Under such circumstances, rare types are at an obvious disadvantage. Discriminatory practices may arise to reinforce these preferences, making fitness strongly frequency-dependent. We are all too familiar with such behavior in our own species, and the burdens suffered by minorities within any society as a consequence.

All of these mechanisms also operate among populations, and even among species. Resistance to a pathogen, for example, is selectively advantageous only in an environment where the pathogen exists. The same gene that provides resistance to malaria also can cause sickle-cell anemia when it is inherited from both parents. Thus, the gene is a seriously disadvantageous one in a malaria-free environment. More generally, the adaptive landscape for any species is linked to the characteristics and abundances of other species; indeed, interacting populations, such as parasites and hosts, may be intertwined in an ongoing coevolutionary dynamic that influences changes in both.

Thus, the evolutionary context for any species includes not only the physical environment but the biological environment as well. As evolution proceeds, it changes that biological environment, altering selection pressures and the adaptive landscape. Indeed, physical and biological environments are so intricately related in their influence on evolution that it is impossible to separate them. Even such a basic adaptation as the geometry of a tree's branches or roots, fundamental for the acquisition of sunlight or other resources, is profoundly influenced by the geometries of other trees. The study of evolution addresses problems in game theory, not ones of static optimization. I turn next, therefore, to consider the ways in which organisms interact, and the implications for the evolution of ecosystem structure and functioning. In this way, we shall begin to explore how patterns of interaction emerge from selective pressures within and among populations.

The Dynamics of Competition

In a world of limited resources, evolutionary success is measured in relative performance. No population can increase forever, and competition shapes the adaptive landscape.

Any environment, whether a forest or a grassland or a lake, is a mosaic of *microenvironments* with different physical and chemical attributes. Organisms may in principle become adapted to virtually any microenvironment, but it remains true that some are inherently more productive than others, more fertile ground for exploitation. Why bother adapting to the less favorable environments when more favorable ones exist? The answer is obvious, and familiar to any sane businessman who opens a store in a remote area rather than in the most populated areas: competition.

At the core, all other things being equal, populated areas are potentially more productive sources of revenue, but total revenue is not what counts when it has to be shared with a large number of competitors. In an imaginary city with no stores, the fitness initially is highest for stores that locate in the most populated areas; this location in turn, however, reduces the marginal fitness gained by expanding such stores, or the fitness available to new stores. Outlying areas begin to look more attractive in comparison. Revisit the analogy of a fitness landscape that is a bit like a huge, soft waterbed, with people strolling on it and trying to climb local peaks. Each step in the direction of higher ground transforms that high ground into low ground, while releasing the points left behind to reach new heights. (The analogy is admittedly not a hydrodynamically perfect one). In the same way, adaptation to specific microenvironments reduces the inherent attractiveness of those microenvironments relative to others and directs attention toward the exploitation of previously neglected resources.

This principle is basic to the evolution of natural communities, and it is the major force for generating and sustaining diversity. The adaptive landscape is always undergoing some change, but to the extent that it settles down at all, it must be composed of peaks or ridges all more or less of the same height, separated by valleys. Were this not true, those sitting on the higher peaks would be expanding at the expense of the others, undermining any semblance of equilibrium. Occasionally new and unexploited environments appear, owing, for example, to climate change or volcanic renewal. Competition is temporarily suppressed, and populations can grow. The opportunities associated with these novel habitats create new and higher blips on the landscapes; these are rapidly found and exploited by evolutionary forces, however, until, of course, they, too, become saturated and the adaptive heights once again are equilibrated.

Microenvironments appear in time as well as in space. Indeed, strategies such as dispersal exploit the interchangeability of temporal and spatial variation, allowing organisms to use movement in space to find niches in time. Successional species in forests provide obvious examples, but it is illuminating as well to consider the adaptive forces that have helped shape the strategies of migrant birds. Wintering in the tropics or subtropics is not just a pleasant way for birds to catch a break between terms; it is necessary to their survival, a way to escape the vicissitudes of winter. Birds do this by migrating, bears hibernate, insects diapause, and plants make dormant seeds. All are playing the same game: wait out the bad times and be ready for the good. What determines when birds return to northern climes from their winter holidays, or when seeds break dormancy, bears awake, or trees flower? Premature reentry into the summer environment carries severe risks if there is still the potential for major freezes, but species that wait too long may lose opportunities for growth. Indeed, and most important, available sites may be limited, and the tardy will find that the ship of opportunity has sailed. So evolution shapes compromises, compromises determined both by physical constraints and by resource sharing.

In the absence of competition, the strategy of choice might seem to be one that would maximize the expected potential reproductive output. February is too early to return to upstate New York because there surely still will be harsh storms. August is too late because insufficient time remains for breeding. So somewhere in between must be best, it would seem, and hence natural selection should lead to the evolution of birds that migrate at just the right time. But discussing natural selection in the absence of competition is like discussing falling objects in the absence of gravity; competition is what makes selection work by choosing those types that make the best use of limited resources. Were winter not a factor for migrating birds, competition would favor earlier migration; as we all know, the early bird gets the worm. In arriving earlier, a bird gains better access to limited breeding sites in exchange for accepting a higher risk from harsh weather. The types that are selected by evolution do not represent optima on a fixed adaptive landscape but rather are the victors in an evolutionary game. These are what John Maynard Smith has termed *evolutionarily stable strategies:* strategies that once established in a population cannot be beaten by any rare mutant strategy.[13] Ritual mating dances in bees provide a case in point. When rare, such behaviors would bring no reward and would surely be costly to the bee; when common, they mandate conformity for those who hope to reproduce. Evolutionarily stable types thus are the best that can be in a world made up mostly of their own type; they may indeed fare poorly under other conditions.

An analogy might be helpful here. If there were only one basketball team in the world, it would probably pay its players very little, if at all. It would have no income stream; few people would pay to see such a team play because with no opponents it could do little but practice. (The Harlem Globetrotters solved this problem in part by becoming pure entertainers and bringing along a token opposition squad on the same payroll.) Things become more interesting, of course, when there are a number of teams representing different cities. People from these cities quickly come to believe that their own worth as humans is somehow tied up with the success of the team that represents them, so they send resources to their favorite team in exchange for tickets to watch them, or clothing that will make others think that they play for the team or own part of it. Better teams usually attract more resources from their fans but cost more to sustain; hence, competition for the best players intensifies, and salaries escalate to decent living wages, such as $20 million per year for talented young high school and college graduates. Strategies have been dramatically changed by competition, and the continual evolution of salaries changes the context in which future salary evolution must take place.

The basketball metaphor is a limited one for evolution because basketball teams do not give birth to offspring teams that share their features. Other examples from the world of business, however, push the analogy a bit further. Chain stores—in effect, the cloning of a successful model store in order to produce similar stores in other locations—change the competitive landscape and drive out mom-and-pop shops that have been successful for generations in the absence of serious competition. As a chain proliferates, it changes the adaptive landscape; if it eventually monopolizes the whole market, it is at a game-theoretical optimum. By selling for less, it makes less profit but is able to dominate the market. This brings to mind the old joke about the store owner who said that with his low prices he lost money on every sale but made it up in volume.

For most sectors of the economy, no single chain completely dominates. Although diversity may decrease over time as chains proliferate, coexistence among these giants is the rule rather than the exception, whether the products of interest are hamburgers, stereos, or sneakers. Each exploits a different microenvironment, and there is no single evolutionarily stable type. Rather, the totality of strategies in the market collectively defines the context that determines the relative successes of types. Competition begets adaptations that provide temporary advantage, proliferate, and then are sustained because they are necessary in an environment where everyone has them. They are, in essence, evolutionarily stable. Airline frequent-flyer programs provide an obvious example of a market feature sustained solely by competition, a fiscal analogue to the elaborate plumage of peacocks. Both are costly to maintain and could be eliminated by collusion, but both are maintained because collusion is not permitted. Every airline would be better off were there no frequent-flyer programs. But if there were none, there would be great temptation for some airline to start one, since it would enjoy an immediate advantage, and others would be forced to follow suit. This is indeed what happened as such programs developed.

In the marketplace, diversity emerges naturally from competition and the benefits of exploiting novel ways of making a living. Such endogenous emergence of diversity also typifies biological evolution. In the case of migrating birds, for example, Yoh Iwasa and I have shown that selection should maintain a diversity of strategies, each making slightly different compromises between physical and biological stresses. Thus, some birds migrate north early in the year, risking winter in order to reduce competition; others come later, preferring to deal with other species rather than harsh environments. This time-sharing may be seen within species, and it may be seen among species. The same multiplicity of strategies applies to germination or flowering times of plants, to emergence times of insects, or to the specialization by plants to different stages of forest successional development. Biodiversity feeds on itself in the sense that it is the changing adaptive landscape associated with the establishment of certain types that makes alternative lifestyles attractive and enhances diversity.

The notion that competition, in concert with other factors, increases diversity rather than decreases it may seem counterintuitive; Wright's adaptive landscape and Fisher's theorem tell a story of a continual erosion of heterogeneity, within the limits set by genetic mechanisms such as mutation and recombination, which restore it. It is frequency dependence, within and among species, that makes the difference between the real world and the mathematical abstractions. If evolution had to deal only with a fixed set of engineering challenges that could be solved by optimization, the world would be a rather boring place, with a depauperate biota. Frequency dependence, however, leads the process of evolution itself continually to open up new opportunities for adaptation, building diversity upon diversity to create the rich flora and fauna that we marvel at.

The rocky intertidal, for example, is streaked by sharply delineated zones defined by the dominant species at different tidal heights; the variegated patterning is so strong that it is easily detected from great distances. Barnacles, as a case in point, are the most evident species in a narrow band above the zone in which the competitively superior mussels dominate, and below another zone occupied primarily by algae. The upper limits to the range of barnacles, like the upper limits for most species in the intertidal, are determined by the physiological limits of the barnacles' ability to deal with dessication; the lower limits are determined by biological factors, in particular, competition by mussels. The reasons are similar to those that structure successional gradients, only here space rather than time provides the gradient. The lower zones, being underwater for a greater period of time, represent relatively stable habitats. In these, colonization of new sites is less important and competitive ability more important. Thus, the competitively dominant species, analogous to the canopy species in forests, are most abundant in the lower zones. Higher up, tidal variation makes environments more ephemeral, favoring shorter-lived species that can colonize rapidly. Furthermore, the upper zones of the intertidal are more stressful, because desiccation is such a dominant factor. In these regions, therefore, selection must first deal with meeting the minimal requirements of the physical regime; competition is a secondary concern.

Barnacles can live throughout the intertidal up to their upper limits and are also found in patches in the lower zones; indeed, they can grow to larger sizes, meaning greater reproductive capacity, in zones below where they most commonly settle. In those lower zones, however, they meet competitive stresses that may overwhelm the other advantages. This story is repeated, *seriatim,* for other species in other zones of the intertidal, leading to a diversity greater than would oth-

FIGURE 6.2 A stretch of the intertidal zone at Tatoosh Island, Washington. Visible upper zones are dominated by algae and barnacles, middle zones by mussels and gooseneck barnacles, and lower zones by macroscopic algae. Photo by Simon Levin.

erwise be possible. It is a form of partitioning repeated in other systems as well, such as tidal salt marshes. There, as in the intertidal, the distributions of species along a gradient of physical stress typically show a pattern of species being limited at one end of their tidal range by physical factors (related to submersion underwater) and at the other end by biological factors (primarily competition). The preferred and observed settling depths for any species in these systems, therefore, represent the same compromise between intrinsic favorability and competitive unfavorability that determines the germination times of plants or the return dates of migrant birds. Instead of there being a single type found in an optimal habitat, biodiversity emerges as the child of frequency dependence within species and competition among species.

The emergence of biodiversity through the workings of competition can take a variety of forms. Competition for resources such as space, light, or nutrients is the most obvious example, but competition for pollinators or for safe havens from predation can work just as effectively. A fascinating example of this involves the evolution of *as-*

pect diversity in prey species as the result of *apostatic selection* by predators. Both words require definition. In religion and politics, of course, "apostasy" refers to the abandonment of one's faith or principles. Predators looking for prey typically form *search images* for certain types, the way a human would search for a contact lens lost in a shag carpet. When the preferred prey item is unavailable, the predator may abandon its search image and look for something else—hence the term "apostasis." Birds hunting moths for dinner, for example, may fix their sights on particular patterns on the Lepidopteran wings; the features that define the search image, be they geometric shapes or colors, are known as the prey's aspect. Because predators switch from less frequent to more frequent types, rarity confers a selective advantage. Birds that have a search image for common types—for example, moths or butterflies with specific spot patterns on their wings—will maintain that image; birds that have an initial search image for rare types, finding little success, will switch their images to other, more common types. Thus, through this mechanism, novelty is rewarded and diversity increased.

Crypsis, which describes the ability of prey to be cryptic by evolving coat patterns and colors that permit them to secrete themselves into their backgrounds, can also lead to diversity as different species evolve adaptations to match different backgrounds. One of the most stunning examples of this is *industrial melanism,* through which color patterns evolve in response to industrial factors. The most famous, but not the only, example is of the peppered moths *(Bison betularia)* in Britain. As H.B.D. Kettlewell showed, moths near industrial centers (where tree trunks were sooty) were much darker than forms of the same moth found elsewhere.[14] The darker forms were cryptic against the sooty background, providing them with protection against predation. Eventually, as areas were cleaned up around Manchester, for example, the lighter forms very rapidly returned to prominence through evolution by natural selection.

Crypsis and apostasis interact to shape patterns of aspect diversity. Crypsis involves external factors, such as the sooty British trees near Manchester; in contrast, apostasis describes a mechanism through which diversity arises from nothing (that is, against a *homogeneous* background) and feeds on itself. As with so many of the other problems discussed in this book, understanding the evolution of aspect diversity therefore requires an understanding of the interplay between extrinsic and intrinsic influences.

Character Displacement

As a variety of examples have already shown, competition for common resources, even very desirable ones, can increase the pressure to exploit alternatives. Traits that govern where larvae settle, when birds migrate, what foods animals gather, or when plants flower or set seed all may shift evolutionarily as the result of competition. Such evolutionary change is known as *character displacement.*[15] The most famous examples come from the work on Darwin's finches in the Galápagos done by David Lack [16] and by Peter and Rosemary Grant.[17]

Darwin's finches are birds of various species, among them ground finches, of the genus *Geospiza,* whose success in life is fundamentally tied to the sizes of their beaks. Through the vagaries of colonization, the various Galápagos Islands have distinct groups of resident species of ground finch. On Marchena, Bindloe and Santa Cruz Islands, for example, there are three species of seed-eating finches: *G. fuliginosa, G. fortis,* and *G. magnirostris.* Daphne Major, where the Grants have done their classic work, and the group of four islands known as Los Hermanos (the Crossmans), in contrast, each support only a single species.

Lack showed that the beak lengths of ground finches differ more when they occupy the same islands than when they exist in isolation; such evolutionary displacement of a feature or character—in this case, beak length—gave rise to the term "character displacement." The beak length of *G. fortis,* which lives on most of the Galápagos Islands but not on Los Hermanos, is intermediate in size where *G. fortis* is alone but larger when *G. fortis* is joined by *G. fuliginosa. G. fuliginosa,* on the other hand, is intermediate on Los Hermanos but smaller when it shares space with the other species. The Grants have gone the crucial next step, demonstrating that beak size has heritable variation that allows the population to change very rapidly in response to environmental stresses. For example, severe drought during one normally wet period on Daphne Major depleted the supply of seeds, especially small ones, making it advantageous to be able to exploit harder seeds; in a very short period of time, stouter bills evolved in *G. fortis,* the lone species present. The situation was reversed under wetter conditions.

These findings addressed adaptation to environmental factors rather than character displacement, but the evidence that beak sizes are so evolutionarily changeable provides strong support for character displacement driven by competition as well. Beak size, like any other trait, involves a trade-off between adaptation to environmental conditions and character displacement to reduce competition. In this, it is no different than traits such as bird migration time or barnacle settling depth. In other work with Dolph Schluter, the Grants extended their research, demonstrating the importance of character displacement among the seed-eating (granivorous) finches.[18]

The phenomenon of character displacement makes it difficult to deduce the importance of competition from observation of overlap in resource use alone. Joseph Connell termed this the "ghost of competition past": if we see little overlap, it may simply be that a history of intense competition forced displacement.[19] Similarly, apparent overlap in resource use does not mean that species are competing; if they seem not to have displaced in character, perhaps some hidden fine-scale specialization (possibly as a result of invisible character displacement) has minimized competition.

The Enhancement of Diversity

The essential challenge of this chapter is to understand how ecological diversity arises through the process of evolution. Life began with very simple unicelled organisms, which evolved into a staggering diversity of more complex multicellular types. The forest, for example, is made up of multitudes of species of trees and shrubs that ply similar trades but have found unique niches that allow them to coexist. How has such complexity arisen from simplicity? How has ecological heterogeneity emerged from homogeneous origins? And what has limited the spread of heterogeneity and diversity?

The challenges cited in the last paragraph are not so different from similar challenges faced in the understanding of galaxy formation, or the development of economic systems, or simply the emergence of spatial differentiation and patterning in developing organisms or

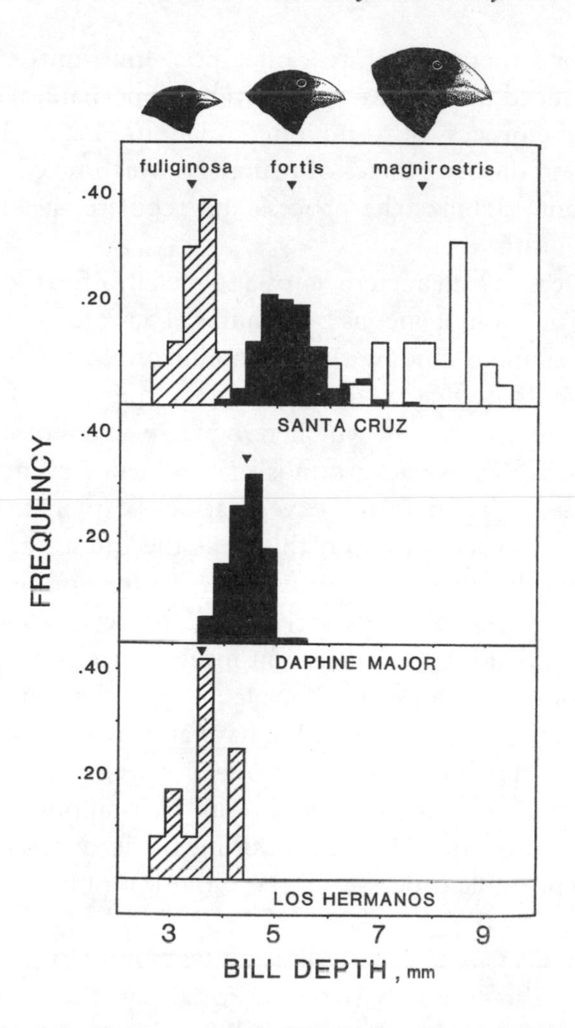

FIGURE 6.3 Frequency Distributions of Beak Depths of Adult Male Finches on Three Galápagos Islands. From P. R. Grant, *Ecology and Evolution of Darwin's Finches*, Princeton: Princeton University Press, 1986, Fig. 15.

chemical reaction systems. These are all problems of *pattern formation* in complex adaptive systems.

In the study of pattern formation, three stages are fundamental: (1) the breaking of uniformity, followed by (2) the enhancement and (3) eventually (possibly) the stabilization of heterogeneity. In a developing organism, for example, all begins from a homogeneous, uniform egg. Uniformity is broken by random fluctuations of chemical concentrations or structural features; once non-uniformity exists, it can become enhanced through local chemical and mechanical interactions, initiating the process of tissue differentiation. Finally, those local interactions and diffusive fluxes of materials (which take longer to become important) stabilize the process and regulate the development of the organism.

These three steps are characteristic of any scenario for the formation of pattern from homogeneous beginnings. They are equally relevant, therefore, to understanding the generation and maintenance of ecological diversity. The first stage addresses the question of why all things are not identical—that is, why heterogeneity ever arose. In the presence of heterogeneous backgrounds that influence events—such as climatic regimes that determine vegetation—explanation is in terms of extrinsic factors (although this begs the question, since it leaves unanswered the question as to why the background is heterogeneous). Against a homogeneous background, however, or for a developing organism that is to a large extent insulated from much environmental variation, the endogenous explanation is that fluctuations that break uniformity are reinforced. We have already seen that a developing embryo, starting from a homogeneous egg and without benefit of a blueprint, self-organizes as a result of local processes that determine the fates of individual cells during tissue differentiation. What are the comparable processes in the evolution of ecological diversity?

For ecological diversity to arise, some rare types must enjoy a selective advantage; if the commonest type were always the most fit, it would displace all others and reign monolithically supreme. In each of the examples presented so far, diversity is enhanced by a form of *negative feedback;* initially favored lifestyles become less desirable as they become more popular, because they overexploit limited resources. Since evolution proceeds by rewarding relative success, the reduced attractiveness of one way of life translates into the enhanced attractiveness of others. As peaks on the adaptive surface become occupied, they sink under the weight of increased competition. No population can for long sustain a growth rate that does much more than just replace its numbers; we are becoming increasingly aware of that Malthusian principle for our own population. In any community, there are always some populations in expansion phases and some on the way out; over the long haul, however, the great majority of species, or of genetic types within a species, are in relative equilibrium. If fitness is measured as the replacement rate per generation, then every surviving type must have a long-term fitness of approximately one, meaning that each individual will on average be replaced by one individual in the next generation. Thus, the adaptive landscape is mostly made up of unoccupied desert of fitness less than one, punctuated by peaks and ridges of fitness equal to one.

The biosphere, of course, is never really at rest, so the view of an adaptive landscape populated by species in equilibrium requires modification. For Darwin, adaptation "was a process of becoming rather than a state of final optimality"[20]; that is, perpetual adaptation does not necessarily translate into optimization. The adaptive landscape is constantly evolving as a result of both extrinsic and intrinsic factors, and frequency dependence and coevolution play fundamental roles in that process.

Stage two in the scenario given earlier involves the enhancement of heterogeneity. In a variety of ways, the establishment of certain species lays the groundwork for the arrival of others, such as symbionts, predators, and parasites. In the colonization of any newly available area, therefore, there may be nearly exponential periods of increase in biodiversity during which colonists set the stage for others to follow and niches radiate. Eventually stage three must set in, since diversity cannot increase without bound.

The same mechanisms are recapitulated over evolutionary time.[21] As Robert Ricklefs writes, "Evolution is self-accelerating in that environmental complexity produced by life forms creates additional opportunities for the evolution of new forms."[22]

Stage three also sets in over evolutionary time. For example, the diversity of marine invertebrates remained relatively constant through the Paleozoic and Mesozoic Eras.[23] Thus, the proliferation of diversity is not a never-ending process, but some balance is ultimately reached between the forces enhancing diversity and those restraining it. Other taxonomic groups show similar slowing down of diversification, even discounting the intrusive presence of *Homo sapiens* as an unprecedented force for extinction.

It is obvious that diversity can increase through exploitation of one species by another, by means of parasitism or predation; the presence of a potential host or prey creates opportunities that did not exist before. Another such mechanism is symbiosis or *mutualism,* in which two species mutually benefit from each other's presence. Lichens, which are coalitions of algae and fungi, are a form of mutualism in which two species cooperate in order to extract nutrients from the environment. Algae perform photosynthesis, converting light energy into organic compounds; fungi provide the nutrients necessary for carrying out synthetic processes by secretion of materials that dissolve the nutrients from bare rock. In this way, the two species become functionally one in their ecological interactions, and organismic complexity emerges.

Mutualism is common in nature, though it is likely that it most typically arises from previously exploitative relationships. *Mycorrhizae,* for example, are soil fungi that derive resources by attaching to roots of plants; they may also, however, transform mineral nutrients, especially nitrogen, into a form that would not otherwise be available to the plant host. The mycorrhizal-plant association thus has evolved into a relationship that is beneficial to both species. Like the lichen, this mutualism depends critically on the tight association between two species, so that, for example, the costs to the plant of giving some of its resources to the fungi are compensated for by the increased availability of nutrients.

Such feedback loops may not require a direct physical link; spatial localization of effects may achieve the same outcome. More generally than mycorrhizal associations, plants and soil microbes compete for the same resources. They live in the same habitats and need many of the same nutrients for growth. On the other hand, plants also depend on microbes in their neighborhood to transform nutrients into forms that the plants can use. Thus, high utilization *(uptake)* of some available nutrients may provide plants with a short-term competitive advantage over their microbial competitors by giving them more resources for growth, but at a long-term cost as other critical nutrients decline in availability with the decline in microbial populations. In marine waters, this feedback loop might be very weak, since local depletion of resources is a short-term cost in an environment in which organisms and nutrients are well mixed. In the soil, however, the effects are longer-lasting, strengthening the advantages of frugality in utilizing available resources; parsimony here in effect translates into the farming of microbes by plants, investing resources that otherwise would be available for a payoff in other and more valuable forms. My former research associate Ann Kinzig, together with John Harte, has investigated these relationships more carefully and demonstrated the essential nature of spatial localization of effects in the evolution of prudent resource use.[24] There is a very profound lesson here for human societies, as I will explore later: where the consequences of one's actions are felt most quickly and most strongly, the motivation for environmentally wise behavior is greatest. To get people to "think globally, but act locally," one really needs to get them to think locally. The most effective ways to do this are to close feedback loops so that the consequences of individual or corporate behaviors rapidly come home to roost.

To some extent, feedback loops tighten on their own as the reward structure reinforces associations. This may occur either in ecological time or in evolutionary time. In our own experiences, we learn quickly who are the most selfish among our acquaintances and to concentrate our generosities on those who will return the favor. The aphorism "You scratch my back and I'll scratch yours" may seem callous, but it guides human actions to a very large extent. From such reliance on reciprocity emerge seemingly altruistic behaviors that are fittingly termed *reciprocal altruism*. These behaviors may extend beyond pairs of individuals, of course, leading to the formation of coalitions that reward, and in turn are held together by, actions that benefit the group. In environmental management, such forms of cooperation have been described by Garrett Hardin as "mutual coercion, mutually agreed upon."[25] It can work, but the larger the coalition, the weaker the sense of belonging and the less effective the feedback loops. Once again, localization of networks of interaction is seen to strengthen the force of selection.

Animal groups provide excellent model systems for studying the benefits of coalitions. For some species, such as our fellow primates or certain equines, troops are sufficiently small that individuals recognize who belongs and who does not. Given such a strong component of recognition and group fidelity, the coalition becomes a unit of long duration, with major implications for both the ecology and evolution of behavior. The beehive presents the prototypical example of a collective that operates as an evolutionary unit; here genetic links serve to hold together somewhat larger assemblages. At the other extreme, however, are the huge assemblages of swarming locusts or of marine planktonic invertebrates, which come together for the transient gains of group living but make no real sacrifice for the benefit of the group. In the latter examples, to which I return in a later chapter, the attraction of individuals toward other individuals produces local clustering and short-term gains in terms of resource acquisition or group defense; but no clearly identified groups arise, and no true group-oriented behavior.

Altruism

Understanding why and when individuals will engage in behaviors that seem to benefit the group at some cost to themselves is central for elucidating the organization of ecological communities, as well as for developing strategies for managing our global commons. The evolution of altruism has been a fascination of evolutionary biologists since Darwin, and its manifestation goes well beyond reciprocal altruism.

For beehives, colonies of Hymenopteran wasps, or mounds of social termites, the role of genetic relatedness in holding groups together suggests that it is the genes that are controlling the action, and that individuals are simply their agents, their interface with the external world. Samuel Butler, in *Life and Habit,* says, "It has, I believe, been often remarked, that a hen is only an egg's way of making another egg."[26] This concept was made operational in J.B.S. Haldane's famous tongue-in-cheek remark that he would lay down his life for two of his siblings, or for eight cousins. The point is that each of us shares about half of our genes with each sibling. The degree of genetic relatedness to our cousins is less. If my mother and my cousin's mother were sisters, for example, they would share (by descent) half of their genes with one another; each of us, furthermore, shares half of our genes with our mothers. So for a particular gene in my possession, there is a probability 1/2 that it has the same ancestry as a corresponding one belonging to my mother (who gave me only half of her genes), a probability 1/4 that it is the same as one belonging to my aunt, and hence a probability 1/8 when extended to my cousin. Haldane was illustrating the point that the same representation of his genes could be achieved in the next generation by saving the lives of eight of his cousins as by saving his own.

William Hamilton, in one of the classics of modern evolutionary theory, worked out the mathematics of altruism, demonstrating why it has arisen most easily in haplodiploid species, largely the bees, wasps, and ants. (As already discussed, in those species, whose males are haploid and whose females are diploid, the birth of males from unfertilized eggs means that brothers are genetically identical, and hence that full sisters—who by definition have the same father—share three-quarters of their genes.)[27] Richard Dawkins carried this a step further in his evocative and provocative book *The Selfish Gene,* where he argues convincingly that the genes are in control.[28] In truth, genes rarely act in isolation from one another; most ecologically important traits are controlled by genes at many different sites or loci on the chromosome. The focus on the gene is appropriate in some situations—it enabled Hamilton to unlock the mysteries of insect societies—but less appropriate in others in which the whole organism, that prototypical coalition of independent genes, is the focus of selection. Nonetheless, the extreme position helps emphasize the fundamental importance of the gene in the dynamics of altruism and cooperation in nature.

Prisoner's Dilemma, which I described in Chapter 2, has been popular among theoreticians because it presents in simple and abstract form the essential dilemma of the evolution of cooperation. A better strategy for the prisoners would be for both to cooperate rather than both defect; mutual cooperation is unstable, however, because playing it exposes one to a cheater who defects from that strategy. *Iterated Prisoner's Dilemma* (IPD), in which the game is played repeatedly, presents a very different situation; furthermore, it is a far better model for the evolution of cooperation because evolution deals with environmental patterns, not with isolated events. In 1979 a political scientist named Robert Axelrod invited some leading game theorists to participate in a tournament in which various strategies for playing Iterated Prisoner's Dilemma would compete.[29] Fifteen contestants sent in candidate programs, which were pitted against one another in a computer. Every strategy engaged in a contest with every other strategy, including its own, and also against a purely random strategy. The winner, surprisingly, was the very simple strategy known as *tit for tat*. In tit for tat, one cooperates initially, and then plays whatever strategy one's opponent played last. Tit for tat was submitted by Anatol Rapoport, a distinguished Canadian game theorist who, not irrelevantly, was also a peace activist.

A tournament is a onetime thing; evolution involves repeated experiences. To study the evolution of cooperation in IPD, Axelrod and John Miller allowed strategies to evolve in a computer competition.[30] Each program was identified with a genotype, with some genes specifying rules for what to play given the recent history of what its opponents had played, and other genes keeping track of that essential history. Initially, selfish strategies began to spread; once they dominated the landscape, however, frequency dependence came into play and reversed the dynamic. Cooperation began to evolve, and eventually it proliferated. Events were not independent of one another because of memory, which provided the reward loop that is essential for the evolution of cooperation.

Repeated play changes the dynamic fundamentally; however, the same result can be obtained if individuals play only once against any opponent but have limited mobility. Studies of Prisoner's Dilemma or similar games among players whose interactions are limited to the individuals in some neighborhood demonstrate quite easily that cooperative strategies can evolve, or that coexistence of strategies may be possible.[31] The explanation is simple: when individuals interact only with others in their own neighborhood, feedback loops are tight and paybacks swift. To simplify the dynamics, suppose that the rules of the game are that when two selfish individuals *(egoists)* play each other, one always dies (chosen by a coin flip), but that when two altruists meet each other, both survive and both reproduce. The new offspring are dispersed into some neighborhood of their parents, not necessarily the same as their game neighborhood. Finally, when an altruist meets an egoist, the selfish individual survives and reproduces and the altruist dies. If the environment can support a limited number of individuals, the death of any individual opens up opportunities for new colonization; this benefits altruists, who must constantly find new places to establish themselves given their competitive inferiority.

Now suppose that, initially, a few altruists are sprinkled onto a landscape populated mainly with egoists. In most places, altruists will be surrounded by egoists and quickly eliminated. In a few places, however, by chance, altruists will find themselves surrounded by their own type; their selfless ways will be quickly rewarded. Clusters of altruists will form, and they will send their offspring into neighboring locales. Such clusters can be invaded by egoists, but they sustain themselves by finding new places to settle. Locally altruists lose to egoists, but globally they may win or at least coexist with their antagonists.[32] Though individuals have no memory, the system develops a memory of its own, through clustering. The mechanism is identical to that which Kinzig and Harte investigated for prudent resource use.

Local Interactions and the Evolution of Good and Evil

My esteemed friend and colleague John Tyler Bonner begins his recent book *Life Cycles* with the sentence, "I have devoted my life to slime molds."[33] Why has so lowly an organism as the *cellular slime mold* fascinated John and many others interested in development and evolution? The slime mold is a model organism; much like a mathematical model, it captures many of the essential paradoxes in a simple and easily studied way. One of the most interesting of its behaviors, indeed, is how it arises in the first place as a coalition of amoebae. The amoebae emerge from encapsulated spores, feed on the bacteria they find, divide, feed and divide, until there is nothing left to eat. To describe what happens next, I can do no better than to quote Bonner, who in his description eloquently betrays his love and admiration for slime:

> Next comes the magic. After a few hours of starvation, these totally independent cells stream into aggregation centers to form sausage-shaped masses of cells, each of which now acts as an organized multicellular organism. . . . After a period of migration whose length depends very much on the conditions of the slug's immediate environment, the slug stops, points up into the air, and slowly transforms itself into a fruiting body consisting of a delicately tapered stalk one or more millimeters high, with a terminal globe of spores at its tip. This wonderful metamorphosis is achieved first by the anterior cells of the slug, which will become the stalk cells.[34]

Now, this has long been a puzzle for those interested in slime molds. Why should evolution permit cells to lead a slug, if the anterior cells are doomed to forgo reproduction, so that their own genes do not make it directly into the next generation? The most likely explanation, again, is geography, and its role in kin recognition. As the slug begins to stream, it picks up recruits from its own neighborhood. Because of the process of cloning that has been taking place, near-neighbors are more likely to be genetically related to the anterior cells than would a randomly chosen individual; thus, a slug most likely involves cells that share genes in common. The altruists, the anterior cells that will become stalk cells, have given up their fitness to benefit their kin; it is once again the genes controlling the action. Altruistic behavior again is rewarded because of the localization of effects.

The examples discussed so far show that what passes for altruism is really enlightened self-interest, whether at the level of the organism or the level of the genes. Behaviors evolve because they have enhanced fitness. I do not mean this as an explanation of human behavior; it is clear that individuals in societies sometimes engage in genuinely altruistic acts that profit neither themselves nor their genes. (Indeed, some altruists are celibate by profession.) While such behaviors may have their origins as adaptations to increase fitness, it is naive to suggest that all aspects of human sociobiology have an adaptationist explanation.[35] Nonetheless, it is useful to understand that evolutionary mechanisms do reinforce altruism, especially when the payback is swift and sure.

It would be misleading to leave the impression that it is only altruism whose evolution is facilitated by the localization of interactions; the same principle applies to any behavior that feeds back to benefit the individual. Symbioses such as those between algae and fungi, or between fungi and plants (through mycorrhizal associations), provide such strong mutual benefit that in some cases the two organisms may be regarded profitably as an ecological unit. Plants and pollinators represent somewhat weaker associations, since a single pollinator may visit many plants, and a single plant may be visited by many potential pollinators; indeed, such visits may cross plant species lines, reducing the effectiveness of the process. This raises a fundamental issue that goes to the heart of understanding how organization arises in natural communities. Pollinators that visit plants of many species may waste pollen, contributing it to lost causes because of the reproductive incompatibility of plant species. This places a strong selective pressure on plants to evolve flower structures that improve efficiency and specialization of pollen transfer. There is also a somewhat weaker selective pressure on the pollinator population, which is in a sense farming flowers. If it persists in inefficient transfers, then its crop will suffer; hence, there is some selective pressure on pollinators to practice what is termed *constancy*, a sort of monogamy at the level of species in which pollinators are faithful to a single species of plant. I say that this is a somewhat weaker selective pressure because the immediate cost to an individual carrying pollen from one species to another is negligible, whereas the cost to the plant is real. Nonetheless, there are mutually reinforcing pressures to enhance constancy, and hence to tighten feedback and reward loops. If we imagine an initial community made up of diverse pollinators visiting many plant species, we see immediately that selection will favor the evolution of mechanisms that enhance constancy, because specialist plants and pollinators alike have a reproductive advantage over generalists: plants in such associations receive only pollen that is of value to them, and pollinators gain from having nourished a flourishing resource. Positive reinforcement leads to the tightening of feedback loops, which further strengthens that reinforcement; structure emerges from initially random associations and strong positive feedback. There are other forces at work, however, in any plant-pollinator community, for example, the immediate loss in fitness to a pollinator that eschews an available plant even though it cannot find its preferred type. Thus, the evolution of degrees of constancy involves complex trade-offs and will vary from community to community.

A fascinating example is provided by the orchids, in which a remarkable diversity of shapes, colors, and smells has evolved in response to the selective advantages afforded by specific pollination systems. These systems can become very specific, as in the association between the orchid *Stanhopea* and its pollinators, the euglossine bees *Eulaema*. Individual species of orchids have evolved fragrances that attract specific euglossine pollinator species,[36] often in one-to-one associations.[37] The specificity of the fragrances, as well as coevolved morphological features of plants and pollinators, tightens the feedback loop by reducing the possibility of interspecific transfer of pollen and cross-fertilization of species.

Regrettably, individual advantage can also be gained, of course, by spiteful or other antagonistic behaviors that reduce the ability of a competitor to secure resources. Like the evolution of mutualism, natural selection for such competitive strategies is facilitated when interactions are localized, because the benefits are realized by the individual rather than having to be shared with others. In general, antagonistic behaviors involve some cost to the individuals that practice them, whether those costs are metabolic or simply expressed in increased risks. Unless there is a realized benefit that outweighs the cost, such behaviors will be selected against. For example, bacteria may produce chemicals that are toxic to their competitors. In particular, some *E. coli* are *colicinogenic;* that is, they produce a substance called colicin that inhibits the success of other bacteria that lack resistance to it.[38] (The production of chemicals that harm others is called *allelopathy.*)

The evolution of these antagonistic behaviors can be studied experimentally. If bacteria are grown in mass culture, in which populations are well mixed, colicinogenic behavior does not evolve because there is no net advantage to compensate for the metabolic costs. In physically structured environments, however, such as on agar plates, allelopathy (poisoning other types) may readily evolve. This work, by Lin Chao and Bruce Levin, represents one of the most beautifully elegant laboratory demonstrations of evolution in action and inaction.[39]

The difference between the two situations that Chao and Levin studied is in the tightness of the feedback loop. In physically structured environments, the individuals affected by a *bacteriocin* (a toxic chemical produced by the bacteria) are in the local neighborhood of the colicinogenic bacteria, providing an immediate competitive advantage to the bacteria that produce the chemical. In mass culture, in which things are mixed up, there is no such advantage. Allelopathy is not restricted to bacteria; it is, for example, a well-established adaptation for competition in plants of the highly structured and highly competitive chaparral plant communities.[40]

Parasites and Hosts, Plants and Herbivores

Parasite-host interactions provide some of the clearest examples of true coevolutionary systems in which evolutionary changes in each species drive changes in the other. Clearly, the fate of the host is affected strongly by the parasite, which saps resources from its host and may ultimately kill it; evolution of resistance to parasites is thus easily understood. The parasite, as well, is dependent on its host, which is its home. If a parasite is too virulent, its host will die too quickly, and the parasite's ability to survive, to reproduce, and to disperse its offspring will be lost. This has led to an acceptance in parasitology that virulent parasites will become less so over evolutionary time, and that, indeed, *commensalism* (peaceful coexistence, in which "parasites" no longer harm individual hosts) will be the end result.

The first part of this story is true, but the latter need not be so. The evolutionary forces that determine a parasite's virulence illustrate clearly a situation in which short-term individual benefits are in conflict with what is best for the group. Only when the success of the group feeds back to affect the individual's fitness on relatively fast time scales can such influences represent important evolutionary forces. In this example, therefore, is a lesson regarding environmental management, in particular concerning when and how individual behaviors may be directed to the greater good. I return to this theme in the last chapter.

A dramatic example of these forces at work is provided by the history of the viral disease myxomatosis, introduced to control rabbit populations in Australia and Europe. The European rabbit, *Oryctolagus cuniculus,* was brought to Australia in 1859 and rapidly reached epidemic proportions. The fabled reproductive potential of rabbits is not an exaggeration, and the twelve pairs of rabbits first released on a ranch in Victoria became an incredible several hundred million within forty years. Rabbits had devastating effects on the landscape, and especially on economically vital pasturelands. Eventually, after several failed attempts at solving the problem, the Australian government imported a virus, myxoma, related to smallpox but harmless to humans. Myxoma had coevolved with South American rabbits, with which it hence coexisted. Introduced to its new host, the European rabbit, the virus, freed of coevolutionary constraints, had a field day. It killed virtually every rabbit it met within fourteen days of contact and had the desired effect of decimating the pesky rabbit population.

The story does not end there. Within a few years, less virulent strains of the virus had displaced more virulent types, providing enough genetic variation in response that rabbits began to evolve resistance to the virus. Commensalism was not the outcome, however; the system has remained somewhere between the two extremes of virulence. Rabbit populations recovered somewhat, and they remain pests today, the subject of heated debate regarding the use of other pathogens for their control. Why did the system not evolve toward commensalism? Evolution in virus populations is a two-stage process. Within a host, virulence is favored, because it increases the virus load an individual host will carry and increases the chance that a *vector* (which carries the disease from host to host), typically a mosquito or a flea, will transmit the virus after it has a meal. On the other hand, the more virulent forms of the virus kill their hosts too rapidly, reducing the chance that an infected host will live long enough to serve as a source for spread. The optimum virulence lies somewhere in the middle of the spectrum. There is no better documented case of coevolution in action, owing to brilliant analytical work by the legendary parasitologist Frank Fenner. The myxoma-rabbit story thus has become a paradigm for what evolutionary biologists term *tight coevolution* (the modifier "tight" here means that two species are involved in very close association).[41] I return shortly to contrast this with *diffuse coevolution,* in which many species interact more loosely.

Parasite-host systems are ideal systems for studying coevolution, because the fate of each species is intimately tied up with that of the other. An individual rabbit houses a group of genetically closely related virus particles, so that the benefits to a gene that conveys reduced virulence are strong. The kin group associated with a particular host, a special case of what David Sloan Wilson calls a *trait group,* provides the clustering and associated strong reward loops that are essential for selection above the level of the individual to become important.[42]

Individual parasite-host systems provide models for understanding coevolution, but most associations are not so specific. The myxoma virus, in effect, jumped from one host rabbit species to another (with the help of humans); this sort of range expansion is typical of the evolutionary history of pathogens and other parasites. As primary relationships favor the restriction of reproductive potential, parasites may find new opportunities in novel host species. In some cases, this may lead to complex life cycles, as for diseases such as schistosomiasis or myxomatosis, in which the parasite must utilize several different host or vector species; more generally, it may simply mean that pathogens maintain the capability of infecting hosts of many different species, adapting to them as they would to any microhabitat. Both examples are familiar to us. We are well aware that many or indeed most of the deadliest diseases that have struck the human population, from plague to smallpox to influenza to AIDS, originated in nonhuman hosts, then adapted to humans when the opportunity appeared. Furthermore, many such diseases can be transmitted only through an intermediate vector, for example, a mosquito, which may have been an original host that has coevolved a commensal relationship with the pathogen.

Similarly, no host species is so fortunate as to need to contend with only a single species of parasite. Any newborn host is an empty island amid a sea of diseases that can potentially colonize it. The human immune system is the result of a diffuse evolutionary response to dealing with myriad pathogen species, including especially the constant barrage of novel types; in turn, it changes the environment for colonization by parasites. Early infections in the life of an individual trigger immune responses that affect which species of pathogen can successfully attack that individual. Thus, the epidemiological history of any population is an experiment in island biogeography, in which pathogens adapted for early colonization (the diseases of the young) affect the later successional development of the individual, and hence the dynamics of the whole community.

It is interesting to ask how such features affect the coevolution of the *etiologic* (causative) agents of disease—the viruses, bacteria, and other microorganisms that plague humanity and our vertebrate relatives. The worldwide range of most disease organisms, however, ensures that any coevolution is very generalized and diffuse. *Farenholz's rule*—the "natural classification of some groups of parasites corresponds with that of their hosts"—has been taken as evidence that these groups of hosts and parasites must have coevolved.[43] The degree to which this is a valid generalization, and at what levels of detail, remains an active area of investigation. There is no doubt—and the examples mentioned above provide evidence—that parasites are not restricted to the host group in which they originally arose, and that this complicates the picture substantially.

Herbivores exploit plants for their resources; predators exploit prey; parasites exploit hosts. In each case, one species derives benefit from another, thereby reducing the success of the victim. The similar asymmetries of these three cases leads to their common classification in many elementary textbooks, but there are fundamental differences. The essential element that distinguishes parasite-host interactions from the other two examples is the tightness of the feedback loops. Whereas a parasite typically lives its entire life on an individual host, an herbivore may move from plant to plant, grazing along the way but not remaining. The death of the host plant is of little importance to the herbivore, and hence selection does not operate on the herbivores to moderate their ways, as it does for parasites whose home is at risk if they are unrestrained in their proliferation. Plant-herbivore coevolution is thus typically diffuse, providing little evidence of strong species-specific evolutionary adaptations. Nonetheless, coevolution is an important mechanism in plant-herbivore systems, shaping the emergence both of defensive chemicals that plants use to poison or deter herbivores and of corresponding detoxification mechanisms in herbivores. These necessarily are fairly generalized responses, however; in the case of plant chemical defenses, for example, there is little direct evidence for tight coevolution between herbivores and plants. Douglas Futuyma, one of the world's experts on plant-insect coevolution, argues that direct connections between the evolutionary divergence of lineages of insects and those of the plants on which they feed are not apparent, and that there is no evidence of a correspondence between the appearance of new species in insects and their host plants.[44] To the extent that coevolution has occurred, it has largely been sequential, with the generalized development of chemical defenses preceding by long periods the evolution of specific adaptations by herbivores to overcome those defenses.[45] There are counterexamples, as in the much more limited associations of vines of the genus *Passiflora* with heliconid butterflies; the general pattern, however, is not only clear but a necessary consequence of the generally weaker feedback loops.[46]

Predator-prey interactions are also diffuse, in general even more so than those between herbivores and plants. Just as humans exploit many resources, so, too, do predators feed on many species. Still, the notion that predators might reduce their exploitation of prey as a long-term strategy for preserving the prey population—that prudent predation might evolve—has been a controversial fascination for ecologists.[47] Prudent resource use by humans, after all, is a central goal of most strategies for environmental management, and the hope has always been that we can find and learn from examples in nature. Such theories, however, have received little support, since they seem to rely on evolution operating at the level of groups, without evidence of mechanisms. Indeed, the objections among evolutionary ecologists have been similar in kind to those directed against the strong forms of Gaia theories.

There is some hope for such a theory, however, through recent demonstrations of how prudent predation might evolve. In his Ph.D. work at the University of Wisconsin, Eric Klopfer used computer models to study the evolution of reduced attack rates by predators.[48] I mention this latter work not just because Eric married my daughter Rachel, but because it once again illustrates the importance of spatial localization of interactions. As demonstrated in the work of Kinzig and Harte, the individual that depletes its local environment too voraciously has no tomorrow to anticipate; more prudent consumption of prey is sound agricultural practice if your future depends on the local prey population.[49] Again, prudent behavior arises because paybacks are swift and localized. The mechanism is the same as that which underlies the evolution of altruism or of many other strategies that we have met in this chapter.[50] I have belabored this simple point, and will continue to do so, because it is one of the central messages of this book.

Coda

Examination of coevolutionary interactions between pairs of interacting species illustrates both the potential and the limitations for the evolution of ecosystem structure. Where feedback loops are tight, strong selective pressures emerge, and these can lead to mechanisms to tighten or reinforce the nature of the feedback. Examples include the evolution of tight mutualisms, such as those provided by lichens or by root-mycorrhizal associations, as well as reduced *virulence* in host-parasite systems. As the perspective broadens to larger numbers of interacting species—for example, plants and herbivores or prey and predators—the linkages become more diffuse, and the evolutionary forces weaker. Some features of communities can still be understood as the result of diffuse coevolution, but these are more likely to have arisen sequentially than as the result of feedbacks in which the preservation of system integrity was at issue. Even in diffuse coevolution, however, spatial localization of effects can lead to strong linkages of actions and effects, and hence to coevolution of general characteristics of the interactions between exploiters and victims.

In this chapter, I have begun to show how evolution shapes the properties of communities. The workings of evolution are clearest at the level of genes and individuals and become fuzzier as we move up the chain of organization—to groups and populations, to interactions between species, and ultimately to ecosystems and the biosphere. Indeed, my central thesis has been that ecosystem structure and dynamics emerge from selection operating at lower levels, and that feedbacks from higher levels are weak because of the individualistic distribution of species.[51]

The emergence of ecosystem structure and function, however, has profound implications for how ecosystems function, and in particular for their resilience in the face of perturbations. In the next chapter, I explore these issues, turning in the following chapter to try to put things together and ask how evolutionary processes at lower levels combine to affect emergent ecosystem properties.

7

ON FORM AND FUNCTION

My desk and its drawers are full of items—both utilitarian and frivolous—that have become assembled with very limited planning by me. Unlike the model of an efficient office, my collection is a product of its history—of the needs of the moment and the vagaries of my travels. The hole punch, the stapler, and the tape dispenser all reflect some degree of intent on my part, as well as continued function on theirs. I bought them because I needed them, and I continue to need them. The collection of ballpoint pens emblazoned with advertisements, mostly for hotels and motels, similarly fulfill an important function, though their acquisition was more accidental than planned. It is a bit harder, however, to justify the world-class collection of pen caps—remainders and reminders of an earlier time when they had an important role in keeping my shirt pockets ink-free, but now long separated from their mates. I keep them against the day when inexplicably all the caps on my working pens will disappear and it will be necessary to press this emergency supply into service. The pen caps at least once had a function. They are thus more easily explained than the diversity of Lucite paperweights, in principle ready to protect piles of papers from nonexistent gusts of wind; or the business cards of people I knew I would never have need to contact; or the motley collection of other trinkets and whatnots accumulated from diverse travels and visitors.

The contents of my desk may not seem to have much to do with ecosystems, but take another look. My office is, indeed, a self-organized system, with me at the center (at least from my perspective). It has more the element of design than do ecological systems, yet it still reflects a huge dose of chance and historical influence. More generally, in fact, my whole house is a museum of memorabilia for which my wife and I have become the curators, simply because we happen to have acquired them at some time in the past; a wax chicken is one of our favorites, reminding us of the friends who gave it to us but serving no other conceivable purpose. Ecosystems also preserve the memories of their pasts. Though many species that they contain are critical to their continuity, others are simply decoration. The hand of historical accident is strong. Even the key species, those that define the character of the system, represent chance colonizations that became reinforced ecologically and evolutionarily, thereby shaping further community development. Other species are like the pen caps, perhaps important at an earlier stage and hanging around now because there is no strong force for their exclusion. Still others are the paperweights: the novelty value that preserves them enhances their own fitness but has little implication for the system as a whole.

In the previous two chapters, we learned how communities become assembled, and how evolutionary forces respond to and shape the players over longer time scales and broader biogeographic regions. What are the implications of this process of self-organization and evolutionary reinforcement on the properties of ecosystems? What structural properties emerge, and how do these influence how systems operate? Where redundancy is evident, is it likely to prove useful, like the spare lightbulbs I stock, or irrelevant, like the pen caps?

Biodiversity and Ecosystem Processes

Investment advisers tailor portfolios to the needs and adventurous tendencies of their clients, balancing high-end potential with risk, both short-term and long-term. Similarly, for ecological communities, these are the key elements in representing system functioning: the average or expected productivity (conversion of sunlight energy into biomass), the stability or predictability of this over time, and the resilience in the face of major environmental perturbations. In this classification, I am using a very narrow definition of stability—low variability in the face of the normal sorts of fluctuations that challenge species.

One of the basic principles of sound investment is diversification, for example, through mutual funds. A balanced portfolio, though it has a lower ceiling for growth, typically exhibits more stable dynamics, reducing the risk of catastrophic collapse. Over longer periods of time, however, unless steps are taken to maintain diversification, diversity will be lost as the faster-growing portfolio components come to represent larger fractions of the capital investment. Both elements come into play in the dynamics of forests and other natural systems. Differences in the growth and reproduction of various tree species lead to shifts in forest composition and a larger representation by the faster growers or reproducers, unless conditions (including competitive pressures) change to favor the slower ones or the continued immigration into the area by less successful types limits local dominance by a few species.

The management of agricultural systems has many of the features of portfolio management, including the importance of the two competing objectives of maximizing yield and maintaining diversity (especially to reduce the risk of catastrophic pest outbreaks, such as led to the great potato famines in Ireland in 1846–1847). Similarly, marine fisheries, which are not managed in the same sense as agriculture but are also heavily exploited by humans, show the benefits of biodiversity. As I discussed in Chapter 5, though particular stocks, such as the Northwest Atlantic haddock, have collapsed, the catches of other species have increased in response, and the total productivity of marine capture fisheries (78 million tons in 1993) is higher now than at any time in history. However, the palates of consumers do not necessarily find one species substitutable for another; many stocks are in trouble, and there is considerable concern for the future of marine fisheries as this diversity is lost.[1]

The link between biodiversity and ecosystem processes is complex, involving numerous contradictory and interacting relationships. Unraveling such intricacies provides the kind of puzzle ecologists enjoy most. Its solution requires the skillful integration of inductive reasoning and careful, controlled experimentation. David Tilman, at Minnesota's Cedar Creek Reserve, has been one of the leaders in these efforts, designing and executing a Herculean set of experiments to illuminate the key issues.[2] Tilman and his collaborators planted species of grasses in a variety of different mixtures, ranging from a few to many species.[3] Biodiversity proved to be fundamentally important to all aspects of the work: the more diverse communities had higher productivity, were more stable over time, and were more resilient in the face of a major drought. This finding, however, tells only part of the story; the truth is in the details, to which I will return shortly.

The Cedar Creek experiments represent an extremely important source of information, but their value would be limited were they not complemented by other, sometimes contrasting, results from diverse studies in real and model ecosystems.[4] Among the most compelling are the studies of John Lawton, Shahid Naeem, and their colleagues at Silwood Park in England.[5] Lawton and Naeem used a carefully controlled set of environmental chambers, termed the *Ecotron,* in which the vagaries of the environment are eliminated: light, temperature, humidity, and rainfall can all be controlled to the specifications of the investigators. Lawton and Naeem were thus able to construct miniature ecological systems, complete with plants; herbivores (insects and snails) to feed on them; parasitoid insects, whose larvae exploit the herbivores; and detritivores (such as earthworms) to clean up junk (like feces and leftover plant and animal parts), making nutrients available again for plants to use. In these chambers, Lawton and Naeem were able to create communities with different levels of diversity—as few as nine species (two plants, three herbivores, a parasitoid, and three detritivores), and as many as thirty-one (sixteen plants, five herbivores, two parasitoids, and eight detritivores)—and thus were able to compare low-diversity with high-diversity systems (see Figure 7.1).

Naeem and Lawton measured a variety of ecosystem processes, including productivity and the flux of carbon dioxide *(community respiration);* both of these clearly increased with diversity. Their interpretation was a simple one: higher-diversity plant communities have more ways to array their leaves, filling space more efficiently and intercepting more light. These experiments again show that diversity can be important to ecosystem functioning. The mechanistic explanation in terms of the capture of light in no way contradicts the claim that, in general, the dynamics of diverse assemblages are dominated by a few species.

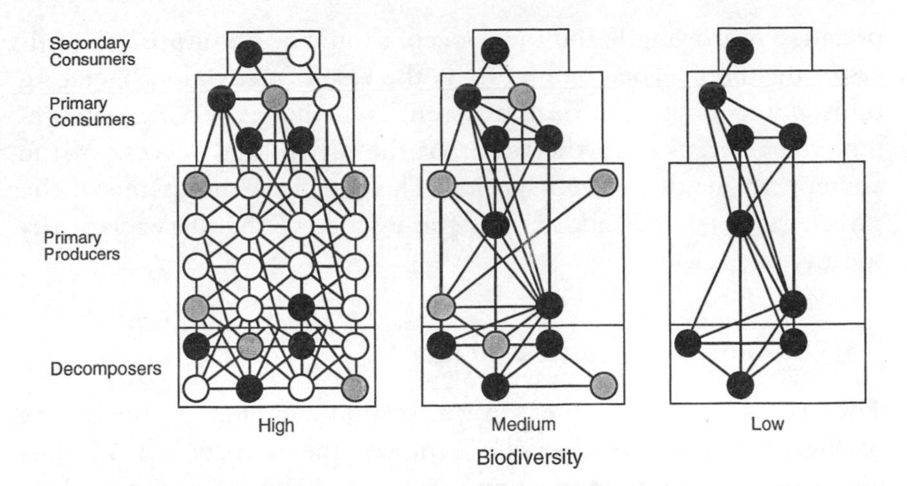

FIGURE 7.1 The Ecotron Ecosystems, Showing Species (circles) and Feeding Links (lines). Modifed from S. Naeem, L. J. Thompson, S. P. Lawler, J. H. Lawton, R. M. Woofin, "Declining Biodiversity Can Alter the Performance of Ecosystems," *Nature* 368 (1994): 734–737. Illustration redrawn by Amy Bordvik.

A variety of competing hypotheses underlie the findings at Cedar Creek and at Silwood Park. Portfolio principles undoubtedly explain some of the results and in particular contribute to the increased stability found at Cedar Creek.[6] Though individual species may fluctuate more when they are mixed with other species (and this is what Tilman and Downing found), the total biomass of all species may be expected to fluctuate less because some species will be at their peaks while others are at their lows.[7] Individual species take up the slack for each other, owing in part to independent statistical fluctuation, and no doubt in part to reduced competitive pressures.

The increased productivity, however, requires further explanation. Indeed, experiments in a wide variety of ecological systems suggest that ecosystem processes typically reflect the influence of a few dominant species, and that diversity itself matters little beyond the identities of those dominants.[8] More diverse assemblages, however, are more likely to contain the most productive species, and this simple idea may explain why more diverse assemblages are often found to be more productive.[9] As the more productive species produce more biomass, they come to represent larger proportions of the community, increasingly dominating overall performance. Support for this interpretation is growing,[10] and the concept should be no surprise to readers of this book. The community is the result of selection operating on its components, not on the system as a whole.[11] Changes in system characteristics are the result of the differential success of the component genotypes and species. This has been the theme of the earlier chapters, and indeed, it is the essence of the Darwinian perspective.

Complexity and Resilience

The studies discussed in the previous section examine only the effects of diversity per se, not *complexity*—the way the diversity in a community is organized into interacting sets of species. Naeem and Lawton studied miniature ecosystems with quite realistic structural organization but took pains to preserve the basic structure while manipulating diversity. Structure makes a difference, however, as is apparent from any examination of why some business organizations are resilient in the face of a changing environment and others fall quickly by the wayside. Organizations that last are those that have the flexibility to adapt to market shifts.

Ecological systems are not random assemblages of parts, but become organized into hierarchies,[12] in which each species interacts strongly with a subset of other species (forming what some call *holons*[13]) and much more weakly with the rest.[14] How this happens is the topic of the next chapter; in this chapter, the message is that this hierarchical organization matters a lot for how ecosystems function. The tropical biologist Egbert Leigh, surely one of the most creative thinkers about such matters, and his colleague Thelma Rowell, invoke Arthur Koestler's parable of two watchmakers, Mekhos and Bios, to illustrate this point.[15] Koestler, in turn, attributes the story to Herbert Simon, who first introduced the watchmakers (whom he named Tempus and Hora) in a 1962 publication.[16] Why Koestler changed the names is not clear; Herb Simon conjectures that Koestler did not like his having mixed Latin and Greek.[17] In Simon's tale, Tempus builds watches part by part; Hora first makes subassemblies of ten parts each, then assembles them into larger units, and so on until the watch is completed. Interruptions are frequent for both watchmakers, undoing whatever progress they are making on the latest units. Tempus finishes nothing, because interruptions always destroy his work. Hora, on the other hand, is never set back as much as ten steps and eventually produces watches. Hierarchical organization localizes damage and provides resilience. More generally, hierarchical structures allow for built-in redundancy, providing another mechanism for resiliency.[18]

Hora has a clear objective, and the modular construction is his way to achieve an end. Ecosystems have no such goals or objectives, and to the degree that hierarchical organization has evolved, it has done so as the result of selection acting on the components, not as a way to confer resilience on the system as a whole. Given that it has arisen, however, it is no less functional for the ecosystem because of its humble origins than if it had been selected for such purposes. The other side of this coin is that self-organized hierarchies may also exhibit features—such as the existence of keystone species—that reduce resilience. We met these power brokers earlier in the book and will visit with them again later in the chapter.

Although Charles Darwin's accounts of his travels gave superb descriptions of feeding relationships—for example, by *holothurians* (sea cucumbers) upon corals—the study of food webs perhaps first began with the work of the major American ecologist Stephen A. Forbes[19] and, simultaneously, the work of an obscure twenty-four-year-old lab assistant in Torino, Italy, Lorenzo Camerano.[20] Most ecologists are well acquainted with the contributions of Forbes; few have ever heard of Camerano, though he probably published the first graph of a food web.[21] The subject did not reach center stage, however, until the famous British ecologist Charles Elton began to publish his long-term studies of the food webs of British woods and to put them into a synthetic framework along with a variety of other similar studies.[22] Elton's food web diagrams not only set the standard for the field but also provided the underpinning for the maps of the flow of energy that are the first step in understanding ecosystems.[23]

Elton was deeply interested in the relationships between community structure and functional properties, such as the resistance to invasion by alien species.[24] Elton's work, and that of Robert MacArthur, stimulated a rich mathematical literature on the possible relationships between system complexity and stability.[25]

Observations of ecosystem structure and dynamics, even over long periods of time, are limited in what they can tell us; controlled manipulations of natural systems were a critical next step in developing an understanding of the relationships between complexity and stability. A major leap in understanding, therefore, came with the work of Joe Connell and Bob Paine on the communities of marine shores.[26]

Connell studied competition between intertidal zone barnacle species, carefully removing one species in order to find out the degree to which doing so allowed the other to do better; this is what ecologists call *competitive release*. As I discussed briefly in the first chapter, Paine carried this experiment a step further, demonstrating that starfish play a critical *keystone* role in structuring intertidal communities. His approach involved careful analysis, logical deduction, the formulation of hypotheses, and, most crucially, critical controlled experimentation that tested the hypotheses. His approach was a model of the scientific method. Having first constructed an Elton-style food web diagram for his study sites (Figure 7.2), Paine had a good understanding of who was eating whom. Furthermore, he knew that the intertidal is substructured into zones—the dominant mussels *(Mytilus californianus)* covered most of the surface in a narrow middle band but were conspicuously absent above and below. The upper limit to the band clearly was set by desiccation: mussels have to be underwater for much of the day, or they dry out and cannot survive. The lower limit required a bit more sleuthing, but Paine knew from his food web studies that mussels were the favorite food of the sea star *Pisaster ochraceous*. Since the sea stars (starfish) are larger than their prey, they have an even greater problem with desiccation; therefore, their upper limit is below that for mussels, leaving a safe band where mussels can survive. (Mussels also do well in subtidal areas, where food is abundant and mussels grow so fast that they can get too large for starfish to attack.)

Paine put two and two together. It is easy to determine the lower limits of mussel beds because the particular mussel species in question is as sessile as a tree once it becomes an adult. Thus, the mussel beds are clearly visible, fixed features of the landscape. Starfish are a bit more problematical for an observer; they move around continuously, albeit at a snail-like pace, and furthermore, they do not blanket the surface like mussels. Their upper range is therefore more variable and more difficult to quantify. Indeed, though it seemed likely that starfish were defining the lower limit of mussel beds,[27] the evidence was circumstantial. An experiment was needed.

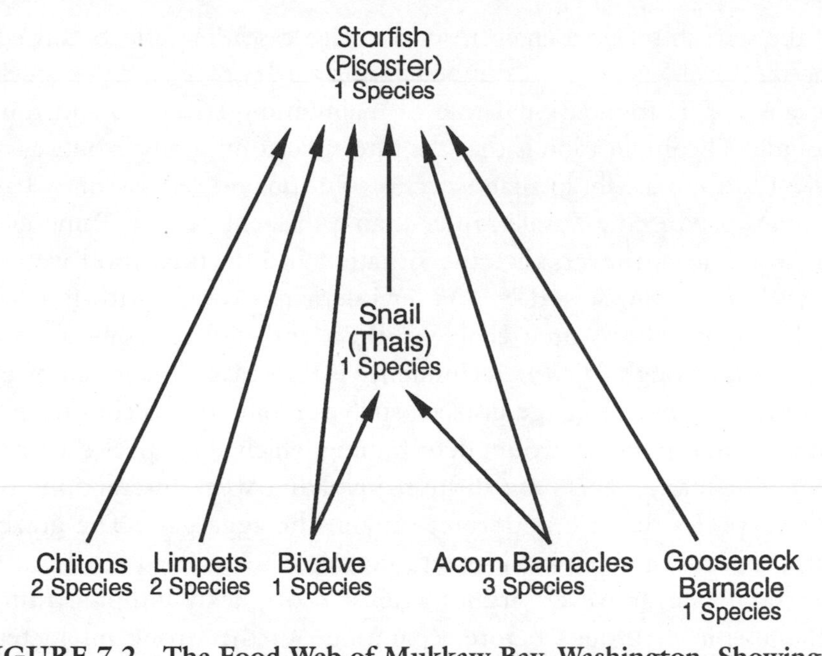

FIGURE 7.2 The Food Web of Mukkaw Bay, Washington, Showing the Flow of Energy. Modified from T. R. Paine, "Food Web Complexity and Species Diversity." ***American Naturalist*** **100 (1996): 65–75. Illustration by Amy Bordvik.**

Paine carried out the necessary controlled experiments on the outer coast of Washington State, originally at Mukkaw Bay, and later on the spectacularly beautiful island of Tatoosh. Laboriously, and with the help of willing volunteers in later years, Paine removed by hand every starfish from a part of the zone below the mussel band, relocating the sea stars to adequate accommodations elsewhere in the intertidal. He repeated this chore in frequent subsequent visits and maintained areas free from starfish. As predicted, mussels began to appear, along with various species of barnacles that were also a part of the sea star's diet. Eventually, mussels took over the entire protected area, eliminating the other competitors for primary space, thereby drastically reducing local diversity; the system had been fundamentally altered by the removal of *Pisaster.*

The results of Paine's experiments were even more dramatic than predicted. Not only did his experiments confirm that the starfish was determining the lower bound of mussels; it also, through this activity, was controlling the entire community structure of the zone. Removal of the starfish set up a chain reaction, with cascading effects through the food web. Paine thus dubbed *Pisaster* a *keystone predator* species because of its foundational role in maintaining structure and functioning. The implication is that the intertidal community is only as resilient as the starfish; in that aspect, evolution and self-assembly have perhaps produced a fragile rather than a resilient system. Paine goes on to argue, however, that the situation in intertidal food webs is much more complicated.[28] *Pisaster* interacts strongly with mussels, and its removal sets up a chain of direct and indirect consequences mediated through *Mytilus;* its influence is thus easily quantified by experimental removal. Large grazers, such as mollusks and chitons, may play a similar role as a group, determining which algal species are present, but many effects are diffuse, involving weak interactions between species that are important only in the aggregate. The grazers form a *guild* or functional group, concepts to which I shall return shortly. Paine, however, argues against too quickly lumping things into functional groups before separating out the strong interacters, lest important information be lost.[29]

Paine's studies revolutionized how people examined intertidal systems and set the standard for the study of coastal ecology. His experimental approach stimulated legions of others to imitate and extend his work.[30] The notion of the keystone obviously had deep implications for the understanding of communities in general, if it were a widespread phenomenon, and people began to look for similar influences in other systems. The most important new example, as discussed earlier, was the California sea otter *(Enhydra luteus)*.

The sea otter was a heavily exploited species on the West Coast of the United States, sought after for its pelts. Eventually, the depredations of humans virtually eliminated the otter from most of the coast, restricting it to a small region near Monterey Bay. When federal regulations prohibited further killing of otters, the species began to expand its range again, with dramatic effects on the near-shore ecosystems. James Estes and John Palmisano, working in Alaska, demonstrated that the otter was a keystone predator in the sense described by Paine, and that its removal would result in fundamental changes in ecosystem structure and functioning.[31] Conversely, the return of the otter to California coastal communities flipped things to the way they used to be, with the cascading implications for fisheries detailed in Chapter 1.

The search for keystones became a major preoccupation in community studies, expanding beyond Paine's original notion of keystone predators to include any species that served as a linchpin in its community.[32] There was no simple prototype—baleen whales and elephants are keystone heavyweights, but rabbits and snails can also play the part.[33] Indeed, one of the best examples was a disease (rinderpest), whose effects on some ungulate species led to cascading effects on other grazers and predators, on grasses and canopy species, and hence on litterfall, fire, and nutrient exchanges.[34] Sometimes, however, it is difficult to identify a single species that fills the keystone role, although keystone groups or guilds play the same role. Paine alluded to this in his discussion of grazing and weak interactions.[35] Predatory ants, for example, form a keystone guild in mountain birch forests, benefiting vegetation by reducing the role of herbivores.[36] Broadening the keystone concept to guilds is an important generalization, extending applicability to a much wider range of ecosystems.

From Functional Groups to the Web of Energy

The notion of the guild has been one of the central concepts in ecology since my Cornell colleague Richard Root introduced it in his classic 1967 paper.[37] The concept of the guild recognizes a ubiquitous and fundamentally important aspect of ecological communities: their organization into functional groups of species that play similar roles, and hence interact strongly.[38]

Root's work has been concerned primarily with birds and insects, but the notion of the guild—which he adopted from the medieval guilds of craftsmen—has much wider relevance. Indeed, the grassland plants that populate Dave Tilman's experimental sites surely form such assemblages, and the competitive trade-offs among strongly competing species explain in part the stability he observed. The same undoubtedly applies to marine fish communities and underlies the constancy seen in total yields while individual stocks fluctuate. The functional approach to ecological communities provides powerful new insights that help us extract order from a morass of detail.

What are "functional groups"? In our own societies, we might immediately think, as did Root, of craft guilds. Plumbers. Carpenters. Doctors. Teachers. Each is a set of people performing similar roles in society. On the other hand, we might prefer to think of doctors, for example, as making up a collection of guilds: internists, pathologists, psychologists, oncologists, and so on. And each of those might be subdivided into even more specific groups. On the other hand, we could "lump" rather than "split," subsuming plumbers and carpenters into a broader category called "skilled craftsmen," or, even more broadly, into the group known as "labor." Clearly, there is no single correct way to do this. The choice of how finely to define things depends on the particular problem at hand. Furthermore, we might choose an entirely different way of grouping people functionally—say, by their degree of power.

We face similar decisions in dealing with the formation of groups in ecological systems. Traditionally, following Linnaeus, species are aggregated into genera, genera into families, and so on.[39] But that is a taxonomic scheme, and it may not be the best way to organize things to represent the functional organization of particular ecosystems. A guild of species then becomes simply any set of species that perform similar roles.[40] So the primary producers might be defined as one functional group, the nitrogen fixers as another, and so on.

In the earlier chapters of this book, I argued that local climatic and other physical and chemical features of the environment determine the broad features of vegetation and the rest of the biota, but that there remains considerable scope for chance and historical dependency in terms of what species actually colonizes a particular habitat. One might guess, for example, that particular kinds of habitats always have the same functional groups but not the same species. Robert Steneck and Megan Dethier, studying coastal marine communities, found just that to be the case.[41] They compared benthic marine algal communities—the species found on the ocean bottom—in environments as diverse as the North Atlantic, the eastern North Pacific, and the Caribbean. The particular species found were quite different in the different environments, so Steneck and Dethier put things into functional categories based on their forms. Some algae are *crustose* (form crusts); these include primarily calcified coralline algae. Other species are termed *turf algae;* these include both filamentous forms and microalgae. Finally, there are the most obvious species, the large *macroalgae* such as the big kelp *Laminaria*. Each of these functional groups could also be subdivided further into minigroups: the crustose algae into coralline and noncoralline, the turf species into microalgae and filaments, the macroalgae into leathery ones and corticated ones (those with an outer layer like a bark or cortex). This way of dividing things was especially useful, because these are the morphological features that define what roles species play in the successional development of the benthic community. And indeed, when Steneck and Dethier used this classification, they found the match that was needed. Provided they knew how nutrient-rich and productive a particular region was, and how important herbivores were, they could predict well the abundances of the various functional groups. Which representatives of those groups were to be found was less predictable; such detail is the "noise" that can obscure the signal.[42] In complementary work, David Rafaelli and Stephen Hall, working in the Ythan Estuary in Scotland, found that most interaction strengths were weak, and that by organizing species into functional groups they could obtain much deeper understanding of dynamics.[43]

The functional group approach helps us to understand evolutionary convergence—that is, the degree to which communities separated by geography, but related in terms of the kinds of environments they inhabit, come to resemble one another. That is not all it does. Since the features chosen define the way organisms function within a system, the functional classification defines how the entire ecosystem functions; of course, it still remains an issue to extrapolate from individual behaviors to those of whole systems, and to explain the importance of particular functional groups. Think of this challenge in terms of the organization of human societies—the issues are the same. A first step in understanding how societies work is to classify people according to their professions. Environment then explains a great deal: it is not surprising that longshoremen and surfboard instructors are found in coastal communities, and coal miners are not. But what determines how big cities will grow, and how many lawyers there will be, and how the stability of the community is influenced by the number of lawyers? These are the essential kinds of challenges in the study of any complex adaptive system.

Some light on these questions can be achieved by the study of food webs. Whereas the basic approach to food web construction involves the identification of which species eat which, ecosystem scientists long have chronicled the flow of energy among compartments of species, such as the primary producers, their consumers, and so on.[44] This is the functional group approach at its most basic. Indeed, Elton pioneered this approach and discovered some important patterns that may be derived immediately from basic thermodynamic arguments. When predators consume prey, a lot of energy is lost through metabolism and heat production; a good rule is that at best 10 percent of the energy at one trophic level may be converted into biomass at the next level. (This is why many argue that our reliance on beef cattle is energy-inefficient: more of the sun's energy can be captured by feeding lower in the food chain.) Thus, it should be expected that the most productivity is to be found at the bottom of the food chain (the primary producers), and that there should be much less production—dramatically less—at each succeeding level in the chain. As a result, much more biomass is locked up at lower than at higher levels, though this calculation is a bit more complicated because of widely varying life spans. Furthermore, since predators typically are much larger than their prey (but note that insects are smaller than the trees they feed upon, and bacteria are smaller than the carcasses they consume), there are far fewer carnivores than herbivores, and so on. All true.

I mentioned earlier the intriguing studies of food webs carried out by Joel Cohen, Stuart Pimm, Peter Yodzis, and others. Cohen, in particular, has reported fascinating commonalities found in food webs when one lumps things into functional groups defined by *trophic level*—the number of food chain steps removed from the sun's energy source. Bob Paine and Gary Polis have taken issue with many of the generalizations; they argue that the analyses are flawed and, more basically, that the world is more complex than simple theories would suggest, but both still argue strongly for the search for organizing theoretical principles married to careful data analysis.[45] Among the most powerful such principles are those that relate the lengths of food chains to productivity: the thermodynamic arguments given above make it clear that the length of chains is limited by the dissipation of energy with each transfer in the chain.[46] It is not surprising, for example, that there are no species that make their living feeding on lions. Even setting aside the rather disagreeable task of capturing and subduing the uncooperative beasts, not enough energy would be available from a particular lion pride to support a population of lion eaters. (Moreover, because the generations of larger animals tend to span greater time periods, the energy consumed is not replenished very rapidly in the form of offspring.)

Because resource demands are so high, animals higher in the food chain (like humans) need to become more catholic in their dining habits and range over wider areas to find food. (The comic-book super-menace Galactus and other interstellar mega-giants must even travel from galaxy to galaxy to find new worlds to consume. It's not really their fault they're so big, though comic books always make them out to be the villains.) Big carnivores must be willing to accept substitutes: one type of food for another, and one locale for another. Humans have perfected this requirement to a tee. Not only are they omnivorous and peripatetic, but they also import other environments by transporting food and other resources from elsewhere to meet the needs of local populations. Evolutionarily, as we shall see, importation loosens feedback loops by eliminating the costs associated with overconsumption, thus reducing selective pressures for prudence and altruism. I return to these themes in the final two chapters.

Closing Thoughts

The structure and dynamics of any ecological community are determined in part by the physical environment and in part by accidents of history, reinforced and frozen in by self-organization. Climate and soil (and the influence of humans) are the principal determinants of land communities; ocean currents and upwellings add a new dimension in the oceans. In general, such features of the physical environment ultimately determine how productive an ecosystem is, and in particular how much of the sun's energy can be captured and used to fuel other trophic levels; primary productivity then sets limits to the size and complexity of the ecosystem that is built on it.[47]

Primary productivity alone does not tell the whole story, as we have seen in earlier chapters. The degree of isolation of a habitat influences colonization, and hence the numbers of species to be found, in accordance with the predictions of island biogeography theory. Historical processes, including environmental stability, also play a role.[48] Of course, competition and other biological interactions are part of the process of self-organization and determine the exact details of the local biota, but many features, such as the total diversity to be found at regional scales, are primarily determined by external factors, such as patterns of temperature and precipitation.

Environment determines primary productivity, and primary productivity influences diversity. But the relationship is complex.[49] Low-productivity regions have very few species. This is not surprising: there is just not enough energy to go around. Where productivity is higher, so is diversity, but only up to a point. When productivity is very high, diversity begins to fall.[50] That is, very high-productivity systems are lower in diversity than are those of intermediate productivity. Theories abound for why this should be true; the most likely is that in very productive environments, so much energy is available that environmental heterogeneity is reduced; thus, a small number of species can dominate everywhere.[51]

An increase in diversity probably also means an increase in complexity and redundancy, with profound implications for the system's total productivity, stability, and resilience in the face of environmental stresses like drought or El Niño events. We have already seen some examples of this, as in Tilman's field experiments, or Lawton and Naeem's Ecotron. Diversity also has well-documented effects on other important biogeochemical properties, such as the amount of organic material in the soil, that are essential for the functioning of the system.[52]

The fundamental challenge remains to determine how structure affects resilience.[53] This is extremely difficult to assess experimentally; work such as Paine's elucidation of the role of keystone species represents the best examples. Those studies show that environmental effects or the major effects on a single species of harvesting strategies would have cascading effects in the ecosystems, because of the way those systems are structured. Similar qualitative changes have also been observed in lakes and other systems but are not universal.[54] Structure matters. Simon's watchmakers hold much of the secret. Cascades of the sort seen in the lake and intertidal systems are much less likely to spread in systems in which compartmentalization reigns—that is, where species are organized into tight clusters of those that interact strongly among themselves and more weakly with members of other clusters.[55]

Understanding how the structure of an ecological system affects its resilience is not so different from understanding how the organization of any system affects its dynamic properties. The movie *Six Degrees of Separation* was built on the theme that we live in a small world—it takes very few steps to get from one person to another—and that fact has implications for everything from the spread of diseases to the stability and resilience of ecosystems.[56] A schoolchild with the flu can infect every other child in the classroom; those children can infect their parents; a parent travels to Washington and meets the president, or to China for a scientific meeting, thereby potentially spreading the disease; the president meets other world leaders and potentially transfers the flu as part of internationalization; those world leaders, as well as the scientists at the meeting in China, may take the flu bug home with them for their families. In a very few steps, nearly everyone could be exposed.

In this way, compartmentalization may increase connectivity, reducing the number of steps between a schoolchild in Peoria and one in Hangzhou, but it also increases the potential that perturbations—such as disease outbreaks—will be contained. The parable of the preceding paragraph would work better for the spread of a joke or an idea than for the spread of flu, because local outbreaks of disease lead to local control measures. A sick child may stay home; high-risk areas may be targets for immunization campaigns. Outbreaks often can be localized and allowed to burn themselves out before spreading to other locales, though obviously major pandemics escape this regional containment. The president probably has infected few world leaders with the flu, though he has probably transmitted many weak jokes to them. In modular structures, local feedback processes may work effectively to suppress perturbations that upset the apple cart.[57] In business organizations, compartmentalization or modularization allows units to be closed out and replaced by others without affecting the overall stability and persistence of the corporation. In businesses, these are design features that may be built in to maximize adaptability. In ecosystems, to the degree that such characteristics exist, they have emerged from Darwinian selection at lower levels.

Resilience and resistance to change are two sides of the same coin. What is desirable in some systems (resilience) is the opponent of modernization in others, preserving social inequities such as class systems, or inefficient economic systems.[58] This chapter makes no judgment about whether resilience is good or bad but simply examines the structural features that contribute to it. In general, these characteristics are "a hierarchy of feedback mechanisms; the maintenance of diversity; options for selection to act upon; and the coupling of stimulus and response in terms of space, time, and organizational scales."[59] These traits combine the general properties of any complex adaptive system—namely, the maintenance of diversity, acted upon by selection—with the special features associated with compartmentalization. The next chapter asks how compartmentalization might arise naturally through self-organization and evolution; in the final chapter, I explore the implications for the management of our environment.

I end this chapter where I began. My desk is, after all, much like the watch shop, or like some ecosystems. When things are chaotic, nothing gets done. But every time things get out of hand, I buy a new desk organizer, with unfilled compartments that can help me restore the flow of work and confine to dark holes the troublesome and irrelevant papers. I have been writing this book in spurts, punctuated by semesters full of other obligations, but its production has been resilient in the face of these interruptions because it is compartmentalized into chapters, and because the hierarchy of desk organizers keeps all in readiness for my next assault.

8

THE ONTOGENY AND EVOLUTION OF ECOSYSTEMS

In the words of Shelley, "Fate, Time, Occasion, Chance and Change? To these All things are subject but eternal Love."[1] We live in an inconstant world, and the ability to adapt to changing conditions is essential to our survival. Success in life requires flexibility and resourcefulness in an unpredictable environment, and evolution rewards characteristics that maintain those traits. Within reason. There is also merit in constancy and in sticking with something proven to work. The key, as in poker, is to know when to sit tight and when to seek new opportunities. Such decisions are repeated over and over again in our daily lives. These range from situations mundane, such as changing one's dinner menu, to those profound, such as changing jobs.

No creature can completely predict its future. Humans, of course, have elaborate devices and formulas to try to reduce uncertainty; so, too, do all other organisms. Even bacteria have ways to determine where conditions are better for them, and then to move there. There are, however, inherent limits to predictability and to what is knowable.[2] In any case, the cost of acquiring information may far exceed the benefits long before the limits to knowledge become imposed. When this happens, it becomes more efficient to let come what may, and to go with the flow. Such principles, of course, impose themselves on our daily lives; more broadly, they influence the evolution of all biological organisms.

A case in point is the common strawberry, *Fragaria virginiana.* Some strawberry plants find themselves in high-light conditions, with an easy source of energy, but others, hidden on the floor of the forest, depend more on chance glimmers of light to fuel the photosynthesis needed for their growth. Given such different conditions, evolution might have produced high-light and low-light specialist strawberries.[3] Instead, it has produced individuals that can change their photosynthetic machinery to suit different environments.[4]

Flexibility in photosynthetic response enables plants to wait to see what their environments are like before deciding what kinds of strawberries to be. Environments change. A plant raised in the forest understory may suddenly find itself in the open when the tree shading it topples. Under such circumstances, it can change its biochemical and photochemical machinery and become a high-light strawberry. This flexibility is termed *plasticity.* Plasticity is important to the plant, but it comes with costs and benefits. For a plant to alter its machinery, energy and time must be expended. The plant's only clue that things have changed is the increased availability of light; if that clue were the result simply of the fluctuating position of leaves due to wind movement and the daily travels of the sun, then a premature response by the strawberry would waste energy and leave the plant unprepared to deal with the soon-to-return low-light environment. The plant must be able to read varying light conditions, gauge their patterns and duration, and decide whether change is for real before making a major shift in its chemistry.

Plants deal with unpredictability at many levels, and evolution has shaped their ways of life accordingly. There is no unique answer to the questions posed by unpredictability, and the diversity of possible solutions becomes translated into the diversity of nature. Deciduous trees shed their leaves each winter, shutting down most operations until the weather becomes more favorable; evergreens, on the other hand, keep their foliage, protecting their leaves against the winter, for example, by shaping them into waxy needles. Annual plants may deal with the vagaries of the environment through dormancy, and some mode of seed (and pollen) dispersal allows all plants to find new opportunities and to average out the unpredictabilities associated with particular locales.

These challenges are faced by plants and animals alike, and solved in similar ways. Food is not smeared like butter evenly over the face of the earth but comes packaged in transient clumps that animals find and consume before moving on to greener pastures. Movement is an essential device for dealing with local unpredictability. But it is not the only mechanism. Diapause and hibernation, the analogues of dormancy, allow animals to last through unfavorable conditions, waiting for better times. Such patterns are also played out on shorter time scales—animals everywhere, from the desert to the oceans, adjust their activity patterns to the daily cycles of light and temperature, confining their energy expenditures to the times of day when rewards are greatest. There are trade-offs, of course; the desert is home to a variety of nocturnal creatures that have evolved the ability to operate in the coolness of the dark.

Physiological and behavioral devices, including plasticity and simple averaging mechanisms, allow organisms to deal with much environmental variability, but the potential is not unlimited. Climates change, resources come and go, and the adaptations that once were adequate no longer suffice. Old genotypes disappear, and new ones replace them; this is evolution at work. Indeed, over longer periods of time, evolution can operate to maintain the very devices—especially mutation and recombination—that allow adaptation within species. Within populations, mutation and recombination maintain the genetic variation on which selection can act; it is the loss of this variation in small populations that poses the greatest threat to endangered species, because it compromises their capacity to respond to an ever-changing environment.

Certainly, diversity and heterogeneity are equally important at higher levels of organization. Functional redundancy within groups of organisms—for example, the plants that capture the sun's energy, or the decomposers that return nutrients to the soil—cushions ecosystems against major collapse in its critical functions. Nevertheless, even though evolution clearly works to maintain flexibility in how organisms respond to changing environments and also may operate to maintain variation within populations, there are no clear routes by which it can maintain that variation at the ecosystem level. That is not to say that such variation does not arise—clearly, there is redundancy and overlap in function among diverse organisms, providing the system a degree of resiliency. It simply means that such diversity emerges from the competitive interplay among individual organisms, rather than having been selected for the benefits that it conveys to the ecosystem as a whole. In the eighteenth century, the great economist Adam Smith argued that an *invisible hand* guides the development of economies, so that society's best interests result from the selfish actions of individuals. The situation is much the same in the ontogeny (development) and evolution of ecosystems: the resiliency that emerges is the result of a reward structure at the level of individual organisms, not at the level of the whole system. And just as neoclassical economics has challenged the assumptions underlying Adam Smith's confidence, so, too, is there no guarantee or even likelihood that the competitive evolutionary marketplace should optimize features at the level of the ecosystem or biosphere.

Ontogeny and Evolution

Ontogeny refers to the development of an individual organism; indeed, it is a stretch to apply it, as I do, to the development of a whole ecosystem. But the term serves to distinguish the purely transformational changes from evolutionary processes that act on populations of organisms and ultimately shape the rules that guide ontogeny.

Chapter 5 of this book dealt with the ontogeny of systems, and Chapter 6 with their evolution. Indeed, it is easy to think of examples where the two are so neatly separable. Scrabble provided a nice case study earlier: the rules that guide the development of the game may become modified over time to reflect more successful outcomes. The natural evolution of games such as chess or checkers provides perhaps better examples: as such games become popular, mutations occur, providing a diversity of similar games for evolution to act on. Selection favors the development of certain lines, based on their popularity, and extinguishes others; historical accident exerts its usual influence. Many of the older board games still exist, in the sense that their rules are known and chronicled in books for game buffs, who delight in playing them; but for most players the allure of novelty soon diminishes and the reasons that game evolution proceeded as it did become clear. The same can be seen in the evolution of field games, such as softball, or of musical instruments and composition. Plucked string instruments were as popular as bowed ones a few centuries ago, but the evolutionary branch that gave us violins and violas eventually dominated classical composition. Thus, evolution acts on the rules of games, or the principles of composition, over long time scales, opting for systems that enjoy the greatest popularity.

Compared to natural systems, however, games and musical instruments and the majority of consumer products are fairly simple in structure, because their components do not evolve. Indeed, many games and tournaments do include a form of selection: chess, for example, one of the most complex and intricate of board games, has a diversity of kinds of pieces, the weaker of which become eliminated during the course of play. However, with a few arcane exceptions (pawn promotion), even chess does not permit a process of renewal. A few games do allow the fittest types to reproduce: the game of Risk, for example, produced by Parker Brothers, reinforces conquering armies with new recruits of the same type. For virtually all games, however, there is no process, such as mutation, by which diversity is constantly renewed.[5] The games we create are thus obviously pale imitations of the complexity to be found in natural systems, in which games exist within games, and in which the process of development is inextricably intertwined with evolution acting at many different levels of organization, on time scales both longer than and shorter than the lifetime of the individual system.

In the development and lifetime of a biological organism, natural selection is played out many times among the cells that make up the organism. For the most part, the competitive processes at the cellular level operate for the greater good of the organism. The immune system, for example, is a remarkably sophisticated mechanism that stimulates the reproduction of antibodies that repel attacks on the body by foreign agents, called antigens. The antibodies, of course, have no idea that they are working for the greater good; Adam Smith's invisible hand is indeed at work, but in ways that have been shaped by evolution operating at the level of the whole organism. Selection at the cellular level is not always such a good thing, however. At times, the immune system goes haywire, as any allergy sufferer or victim of autoimmune disease knows. Another familiar example, cancer, represents the unchecked growth of certain types of cells at the expense of others, ultimately with dire consequences for the organism.

There are a number of lessons to be learned from the previous paragraph. The development of a single organism reflects evolutionary processes operating at scales both much longer than the lifetime of an organism and much shorter, as well as many in between. The operation of the immune system relies on rapid scale selection and evolution; the structure of the system has been shaped by evolution over longer and slower scales. More generally, longer-term processes reinforce mechanisms that enhance the fitness of organisms; in this way, they shape the rules by which the more rapid evolutionary processes operate, for example, by guiding the evolution of the vertebrate immune system. Those shorter-term processes, however, retain much scope for independence and do not necessarily serve the greater good of the organism, as illustrated by the spread of malignancies. As one moves up the organizational scale—to the population, the community, and the ecosystem—all of these considerations remain valid, but their relative importance changes. At the ecosystem level, for example, or for the biosphere, longer-term evolutionary processes have at most a weak relationship to the fitness of the whole; instead, they reflect emergence from more local processes. There are no evolved mechanisms to deal with the foreign invaders—whether the introduced species that distort natural processes, or the native species, such as *Homo sapiens,* whose growth and impacts are unrestrained. This is why it is so important for us to understand what the natural processes are that guide resiliency, and how to sustain them.

Self-organized Criticality

Imagine that you are among the first to arrive at what will be a very large cocktail party. Only a few people are present; they are all having a lively discussion, which you join. As others begin to arrive, they, too, join the discussion, but soon it becomes too large to maintain and breaks apart into two groups. Eventually, others show up, and the two groups become many—some large, but the majority small. Discussions are transient, and groups split and re-form, split and re-form. No group lasts long, but the distribution of groups retains its character—a few large and many small—for hours, in a state of dynamic equilibrium.

Virtually all mobile organisms, from bacteria to elephants and whales, also organize themselves into groups, and very commonly into similarly dynamic cocktail parties. Swarms of locusts, herds of wildebeest, and schools of fish all exhibit such patterns, arising from continual processes of breaking and joining. The simplest of mathematical models can reproduce the observed patterns, giving rise at times to "power law" distributions in which the frequency of groups of a particular size declines as a power of the group size.[6]

This situation is reminiscent of what the physicist Per Bak termed *self-organized criticality,* a notion that has been one of the most seductive ones in the study of complex adaptive systems. Bak invites only grains of sand to his cocktail party, however, dripping them onto a plate from a fixed position above it (Figure 8.1). Initially, the grains of sand form a single and ever-increasing pile. As the pile gets larger, however, it becomes less stable. Avalanches begin to occur as grains tumble down the sides and gain adherents. Unlike the cocktail party, the sand never forms more than one cluster, but the distribution of avalanches itself fills out a power law distribution: there are many small avalanches and a few large ones. The term "criticality" here is borrowed from solid-state physics to describe a situation in which small perturbations at one point have influence virtually everywhere in a medium, triggering major transformations. The most familiar example of criticality is what happens to water as it is cooled, becoming ice, or heated, becoming steam. At transitions, the material is in a *critical state,* where, for example, crystallization rapidly spreads. The difference is that this is criticality controlled by the changing temperature; it is transient and gives way to the solid or gaseous state as the temperature continues to change. Not so with self-organized criticality, which is a situation that sustains itself by feeding off the avalanches it creates.

Bak, imbued with the unrestrained enthusiasm that theory can confer, sees self-organized criticality at every turn.[7] It represents, for him, the unavoidable outcome of the self-organization of a system of many interacting entities. Ecosystems are perfect candidates for such theory, in Bak's view, and he draws encouragement from a glance at the evolutionary record and at the distributions of "extinction

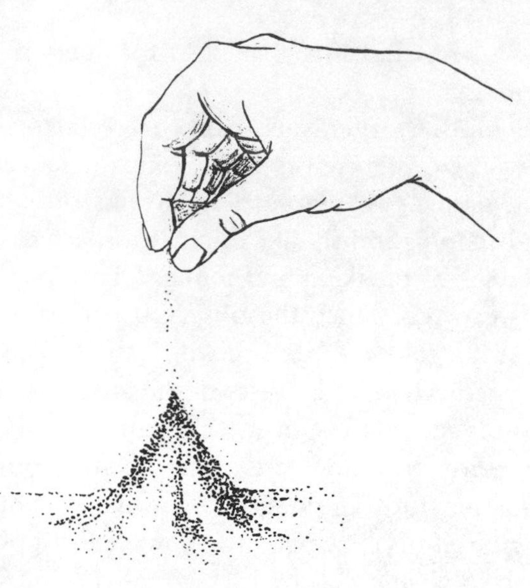

FIGURE 8.1 Creating a Sandpile. Modified from Elaine Wiesenfeld's drawing in P. Bak, *How Nature Works.* Illustration by Leila Hadj-Chikh.

avalanches."[8] In Bak's view, ecosystems organize themselves into critical states in which continual extinctions send shock waves through the systems, triggering other extinctions and reorganization and driving the system to a dynamic equilibrium of turnover and replacement. Over evolutionary time, these events explain—for Bak—the existence of "punctuated equilibria," the long periods of stasis punctuated by major change that Gould and Eldredge claim dot the fossil record.[9] Over ecological time, they are suggestive of MacArthur and Wilson's discussion of island biota, but carried to an extreme.[10] Recall that, in the theory of island biogeography, islands are viewed as being in a dynamic balance between colonization and local extinction. This is a fairly controlled process, however, in which turnover events involve few species. Communities are not really ever saturated, because rates of colonization are restricted by the islands' isolation from the mainland source of species. However, with Robert May, now chief scientific adviser of Great Britain, MacArthur extended these concepts to terrestrial communities that experienced much higher colonization rates and therefore reached saturation.[11] These communities were much like Bak's sandpiles: avalanches of species replacements could be engendered by the continual dripping of species onto an already full plate. There is, therefore, at least some theoretical support for Bak's perspective.

Bak views self-organized criticality in an appropriately nonteleological way. There is no guiding force, no force of evolution driven by what is best for the ecosystem. Curiously, Bak sees self-organized criticality as "the general, underlying theory for the Gaia hypothesis," dismissing opposition to Gaia as the Luddite response of reductionists.[12] But Bak's view of Gaia looks less than searchingly at the maintenance of critical flows and processes and is the antithesis of the view that the biosphere has evolved features because its inhabitants need them.

Bak also takes pains to distinguish criticality from the edge of chaos,[13] and in particular to distance himself from the suggestion that evolution drives systems to the edge of chaos because it is good for them.[14] The edge of chaos is, in its vulgar usage, the boundary between order and disorder—not so ordered that exploration and innovation are impossible, and not so disordered that information gained cannot be put to good use. Most of us, to some degree, must live our lives at that boundary, willing to take a few chances at new things, but unwilling to throw all caution to the winds. Cellular networks in our body, Stu Kauffman reasons, also have been selected to operate at the edge of chaos (here meaning disorder). Why not, therefore, ecosystems? If ecosystems were at such an edge, Kauffman argues, they would be in the optimal condition for adapting to changing environmental conditions. "Over evolutionary time, ecosystems may self-tune to a transition regime between order and chaos, maximizing fitness, minimizing the average extinction rate, yielding small and large avalanches of extinctions that ripple or crash through the ecosystem."[15] There are lots of problems with this claim.

First of all, it is really an untestable concept: there is no sensible measure of the fitness of an ecosystem, given that ecosystems do not reproduce and that their identities are ephemeral in space and time. Second, there is no mechanism operating at the ecosystem level to hone its properties to any set of criteria; what happens, happens, without top-down control. Per Bak's idea is a simple one that rejects any guiding hand: developing ecosystems—as prototypical large, self-organizing systems—continue to grow in complexity and connectedness until they reach criticality, at which point individual species fluctuations drive cascades of variation and extinction of predictable statistical character. Here, the continual generation of novelty through extinction cascades, and the continual loss of species through those very extinction events creates a dynamic and self-sustaining balance. Nowhere is there the suggestion in this that extinction rates have been minimized—quite the contrary. Nowhere is there the suggestion that evolution has driven ecosystems to criticality for their own good—quite the contrary. The ecosystem, in Bak's view, finds self-organized criticality the way a ball finds its resting place at the bottom of a basin—passively, and controlled by natural laws. Bak's claim is, in itself, an attractive hypothesis that suggests a variety of theoretical and empirical explorations. Whether it is true or not remains an open question; my own conclusion is that it contains a kernel of truth but is incomplete in failing adequately to account for evolutionary forces that operate at lower levels of organization.[16]

Self-organized criticality clearly is a feature of many developing systems. It describes a situation in which the effects of small disturbances need not remain contained but may spread rapidly because systems have reached a high degree of connectedness and correlation. When such effects are present, there are no characteristic scales at which events are correlated—they are correlated at all scales, giving rise to patterns of self-similarity and *fractal* landscapes.[17] Linda Buttel and I found such patterns arising from our simple models of forest community development,[18] and others have noted similar features emerging from the analysis both of model outputs and data.[19] Self-organized criticality is not likely to be a universal feature, however. Where strong external disturbances—including the depredations of keystone predators—are important, competitive interactions among species become localized and the interconnectedness of competitors is reduced. Surprisingly, the same effect may result from very strong competitive interactions, which force species to become spatially separated in their distributions.[20] When interconnectedness is reduced, disturbances are contained—avalanches are avoided. My view is that in nature one should see a wide range of kinds of communities—some near self-organized criticality, some as modularized as the watchmaker Hora's workshop, and the great majority in between. It remains to understand why we should see this diversity, and how environmental conditions determine what we should see where. For this, we need an evolutionary perspective.

Evolution

Self-organized criticality is about ontogeny—that is, about the development of individual systems. Though it deals with competition, its evolutionary component is weak at best. Ecosystems in the critical state are characterized by high turnover and heterogeneity; they provide the variability that can foster adaptation to changing environmental conditions and, indeed, endogenously generate much of that environmental change through cascades of collapse and rebirth. Such features thus feed on themselves, maintaining some of the gross properties of the system but exhibiting high turnover in species. Whether one should view such systems as resilient is hence a matter of perspective, determined to a large extent by how stable biogeochemical cycling and other system processes are in the face of perturbation.

No species can exist alone; hence, evolutionary forces must operate to increase certain linkages. Herbivores and predators must develop the capabilities to find and handle certain foods at the expense of others; this is called specialization. The other side of the coin is that complete specialization makes an exploiter highly susceptible to fluctuations in the availability of its resource or to the evolution of defensive mechanisms—tough skins, thorns, or poisons, for example—that thwart the consumer. Thus, despite the obvious advantages of specialization, there also are evolutionary pressures on species to spread the risk, effectively diversifying their portfolios. This is called generalization. Specialization and generalization define the extremes of a spectrum. Indeed, much of evolutionary theory is about the resolution of the trade-offs between these two powerful influences.

The push and pull that defines the degree of specialization of herbivores and predators, as well as of pollinators and parasites, is also manifest in other types of interactions—for example, in the evolution of the interdependence that comes with coalitions and mutualisms. The end result is the evolution of modularity, in which species interact strongly with a limited number of other species and more weakly with the rest of the ecological community. My conjecture, thus, is that evolutionary forces operating at the level of individual organisms lead to modular and hierarchical organization and move ecosystems away from self-organized criticality. I would further conjecture that evidence of self-organized criticality is still to be found at intermediate spatial scales for many systems, but we are far from a general theory as to when that should be so.

It is impossible to understand the mechanisms that have led to the patterns we see in nature without understanding how and where evolution has operated to produce those patterns, or indeed whether it has had any influence. Similar patterns may arise in different systems for quite different reasons, and consideration of pattern alone tells us little about process. The polygonal designs found on the coats of giraffes, for example, resemble structures seen on the frozen tundra or in thin layers of fluid heated from below. Only in the first case, however, has evolution played a role, guiding chemical processes in development so that the signature coat marking always appears, perhaps the result of sexual selection to enhance attractiveness to mates. The tundra polygons arise from a fundamentally different process and rely on just the right environmental conditions; the development of polygons on animal coats is carefully regulated and buffered against environmental fluctuations.

Analogous comments can be made about the superficially similar branching systems of river basins, the bronchial architecture of our lungs, and the distinct geometries of tree species. The sprawling arborescence of a river basin has not arisen as the result of natural selection and thus has a character that clearly distinguishes it from the other examples. The structures of bronchial branching trees—which have evolved to improve the ability of the organism to exchange and redistribute gases—are quite different from those of trees, which furthermore show much variation among themselves. Trees have evolved structures not only to maximize their ability to capture light but also to shade out competitors. Most forcefully, this illustrates that the properties of ecological communities emerge from the selfish evolution of potential antagonists rather than from a cooperative process among members of a monopolistic trust. What is good for the ecosystem is neither well defined nor relevant evolutionarily.

Bottoms Up

The primary unit of selection is the individual. What constitutes an individual, of course, is a bit complicated. For some, it is the gene, for others the whole organism. A few billion years ago, the only individuals were single cells. Because there are some advantages to becoming larger than a single cell, a great evolutionary advance occurred when some dividing cells stuck together, producing the first multicelled organisms. Other multicellular organisms formed simply by aggregation of similar types of cells that did not come from the same parent; the cellular slime molds provide a prototypical example.[21] Multicellularity changed the nature of the game, allowing not only increase in size but also differentiation in function. The various cells in our body, though initially identical, become differentiated and specialized—for example, into heart, lungs, or skin—during the developmental process. Because all of the cells in our body have the same genetic heritage, it is natural to think of humans (or chimpanzees or roses) as individual genetic units; nonetheless, there is competition among the fragments of DNA that make up each organism, and hence there is evolution below the level of the organism.

Even at the lowest levels of biological organization, however, there are some hints that the definition of an individual is not so simple. As discussed earlier, much of the genetic information that determines the fate of a single bacterium is carried on bits of DNA called plasmids, which can be exchanged freely from bacterium to bacterium—even from one species to another. Recall that such important properties as antibiotic resistance are typically carried on plasmids rather than on DNA found on the bacteria's own chromosomes. Thus, a bacterial cell without a critical plasmid is a very different organism. What is the individual here? Is it the bacterial host, in which the plasmid is a guest? Or do host and plasmid together form an individual? There is no single correct answer; it is all a matter of perspective and convention. Similar problems concern tight mutualistic associations of species, such as the lichens formed by partnerships between algae and fungi. Is the lichen an organism, or is it best viewed as a coalition? Again, there is no single correct answer.

The complications introduced in the last paragraph might seem to arise because they involve associations among different genetic types—after all, the plasmid's genes are fundamentally different from the genes found on bacterial chromosomes. However, all individuals, including especially humans, are coalitions of large numbers of distinct genes that are united in common purpose—because they are transmitted as a single unit during reproduction—but nonetheless in competition. Indeed, because plasmids are so easily exchanged between individuals, it is easier to think of them as independent units than it is to view our own genetic material as such; this is a distinction, however, of degree rather than of kind.

The cellular slime molds introduce a different sort of problem, since they arise by aggregation of genetically dissimilar units. The slime mold is composed of social amoebae, which come together to form coalitions called slugs; these move through the environment, gaining adherents to the cause. Eventually, the slugs stand on end, forming fruiting bodies. Only a few cells reproduce, producing spores that seed the next generation. The other members of the slug—stalk cells—have sacrificed their own fitness to the greater good. Why? This represents one of the most fascinating puzzles in evolutionary biology, but the answer can only be that the members of a slug—because of the process by which it forms—are closely related genetically.

Cellular slime molds may represent one of the simplest examples of apparent altruism, if not *the* simplest, but we surely know others. The beehive is perhaps the most familiar, a partnership among genetically similar but distinct genotypes. It is not so different from the slime mold in that all females but the queen forgo reproduction for the good of the hive. Like a multicellular organism formed by aggregation, the hive has an economy that thrives by the differentiation of functions. Termite nests exhibit similar properties, bringing distinct individuals together to form a functioning society. Is the beehive or termite's nest a single organism? However we resolve this problem in semantics, what matters is that the hive represents a coherent evolutionary unit above the level of the individual.

As we move up the ladder of complexity, not only the definitions but also our understanding of the processes become more problematical. Ant societies are highly differentiated collectives of closely related organisms. As units, they have many of the features of single organisms because they maintain their integrity over time, but they still represent associations of individuals in evolutionary competition. Human societies share some of the same features: they represent consortia of individuals who rely on one another. As such, there is differentiation of function—some of us are carpenters, some teachers, some doctors—and consequent interdependence. But the members of a human society are not as closely related as the members of a beehive or an ant colony, and membership in a human society hence tends to be more ephemeral. The human society does not have the continual integrity to function as an evolutionary unit, though the organization of people into societies clearly has deeply influenced the evolution of human culture.[22]

I prefer to view evolution from the bottom up, starting from the individual cell. This is an unabashedly reductionistic perspective, but it mirrors what has happened over evolutionary time. Single cells, through a variety of processes, became organized into multicellular organisms, allowing advantages of size and differentiation of function. Multicellularity arose many times independently, and in a variety of different ways. This suggests that it must be a pretty good idea. In turn, multicellular organisms, like bees, have become organized into superorganisms, like hives, that represent natural evolutionary units. In diploid organisms such as humans, coalitions also form. These tend to be more transient and hence not proper evolutionary units, though they are profoundly important in influencing the evolutionary process. The tighter the genetic relatedness, the stronger the evolutionary force. Hence, families develop strong bonds, and many of our most fundamental behavioral tendencies can be traced to the preservation and protection of the family unit. Religious groups and nations obviously represent somewhat looser assemblages but still involve collections of individuals who are more closely related to each other than they are to humanity at large. In this lies the explanation for many attitudes and practices, including racism and xenophobia, that superficially seem to benefit the group at the expense of others.

Relatedness is fundamental, because altruistic behavior toward relatives enhances the propagation of shared genes. Benefiting a relative benefits oneself, or at least benefits one's genes. But genetic relatedness is not the only way to gain such payback. Coalitions are perhaps most tightly held together when they involve family members, but nonrelated individuals also can form stable, long-term partnerships. These may be held together by rapid reinforcement through reciprocal altruism or through the development of societies with standards of conduct and accepted systems for penalizing those who violate the rules of behavior. Such systems of governance have coevolved with genetically based features but surely reflect a form of evolution above the level of the individual.[23] Their evolutionary resilience takes the form of a resistance to change that freezes the bad in with the good, retaining especially forms of discrimination and intolerance that have evolved because they restrain deviation from the norm and keep in check those who would wander. They also, however, maintain cultural practices that enable societies to function effectively.

Such evolutionary trends have become so ingrained that they control behavior even when the direct reward and punishment loops are removed. Herbert Simon argues that this explains what would otherwise be a puzzle: why workers in big companies work harder than would seem to be justified by the direct costs and benefits involved.[24]

The key element in the evolution of traits such as altruism, or more generally selfless behavior for the greater good, is a tight feedback loop that rewards such behavior. As the last example shows, the feedback loop does not always have to exist, but it must have been a presence at least during the initial evolution, when it became established and frozen in. Such feedback can come through relatedness, reciprocal altruism, binding agreements—or simply through spatial or temporal localization. As discussed earlier, cooperative behavior can evolve much more easily if individuals interact primarily with near neighbors. In these cases, explicit or implicit memory is present, feedback loops are tightened, and cooperation is rewarded. Unfortunately, the same can be said about spiteful behavior, which also evolves much more easily when individuals interact only with near neighbors, as in the case of allelopathy discussed earlier. The reason is the same: if interactions are local, the reward structure is direct.

Given that the localization of interactions speeds evolution, it is not surprising that selection also influences the degree of localization. If individuals can gain an advantage from cooperation, but only with near neighbors, those individuals who interact only with their nearest neighbors and cooperate with them will gain a selective advantage. Evolution thus favors the development of strong family groups as well as of hierarchically organized societies—the small worlds of Watts and Strogatz discussed in the last chapter.

Of course, there are some opposing trends: very tight interactions increase competition and may otherwise disfavor the weaker individuals in asymmetric associations. What we should expect to observe in nature is therefore a balancing of these influences and intermediate degrees of clustering and association; this is, indeed, what we see, at a variety of levels. Wildebeest in herds, or fish in schools, are attracted to other individuals from a distance, but they are repelled at close quarters, keeping some minimal distance from others. Similarly, members of human groups bond together for communal activities but retain the need for some privacy and personal space.

The advantages of clustering are that individuals join together to form collectives for mutual benefit. An individual collective, like a multicellular organism, achieves certain advantages of size—protection against natural enemies, access to mates, ease of finding resources, and the ability to benefit from differentiation of function among the members of the group. More developed societies have fewer jacks-of-all-trades and more specialists. They can afford, support, and profit from a wealth of specialized professions—performers, insurance salesmen, investment brokers, and psychiatrists—that could not be sustained in less differentiated societies. No visible hand specifies exactly how many types there will be, and how many of each type, but the invisible hand of the marketplace encourages diversification and determines community structure.

Similarly, natural communities develop patterns of differentiation and specialization that reflect their ecological and evolutionary stability. In temperate climes, environmental variability is a constant challenge, and evolution has favored plasticity and other devices that allow survival under a variety of conditions; this suggests stronger selection for generalists (jacks-of-all-trades) than in the more constant and stable tropics. Tropical forests, for example, tend to have a higher diversity of types, specialists that can exist only under a more restricted set of conditions. No visible hand specifies exactly how many types there will be, and how many of each type, but the invisible hand of the ecological marketplace encourages diversification and determines community structure.

Coda

The structure and functioning of ecological communities, and of the entire biosphere, reflect the complex interplay between developmental (ontogenetic) and evolutionary processes occurring at a wide spectrum of scales. A perspective drawn from one scale alone is bound to mislead; to understand and explain observed trends, we must be able to integrate the effects of a rich panoply of influences.

Many large, self-organizing systems naturally attain a state of self-sustaining criticality in which continual collapses and innovation reach a dynamic balance; the archetypal model is the sandpile, unceasingly refreshed with new grains of sand dripped from above. The theorist Per Bak and others, inspired by the simplicity of the sandpile model and the paucity of assumptions that underlie it, view self-organized criticality as necessarily a very general phenomenon. In particular, they argue that ecosystems should organize themselves into critical states, and that such forces underlie observed patterns of extinction and punctuated equilibrium.

There is undoubtedly much truth in this view, but it is too simple. In particular, it ignores the role of selection at the diversity of levels on which it occurs. Self-organized criticality implies an uncertain environment for all species; what resiliency occurs is manifest only in broad statistical features, such as the mean numbers of species to be found in an ecosystem, or the rates of extinction. Whether this is good for the ecosystem—an essentially meaningless concept—it is bad news for individual genotypes, which would find themselves in a world uncertain beyond bound. Organisms are constantly challenged by environmental uncertainty but have evolved a remarkable diversity of mechanisms to reduce it, or at least to minimize its consequences. These include behavioral and physiological plasticity; averaging mechanisms such as dispersal, dormancy, and iteroparity (multiple reproduction over a lifetime); and a variety of schemes for gaining information about the environment and learning to utilize it. On somewhat longer time scales, evolution has operated on the rules of the game themselves, determining mutation rates and endorsing sex and recombination as ways to maintain genetic flexibility in the face of a changing environment. The remarkable ability of some viruses and bacteria to change rapidly and elude the immune system or medical practice is a particularly frustrating reminder; evolution clearly favors groups of pathogens that remain moving targets for antagonists. Thus, environmental fluctuation—at least at a broad range of scales—rather than being just a force for extinction, has been a force for adaptation and diversification.

Evolutionary pressures to reduce the consequences of uncertainty do not stop at the level of the single species. Mutualisms and other tight linkages among small numbers of species evolve to provide the participants buffering against environmental vagaries. Ecosystems thus become assembled not into sandpiles but into modularly organized webs of interaction in which clusters of interacting species to some extent insulate themselves from the collapses of other clusters. Individual clusters, since they interact little with one another, naturally exhibit some features of convergent evolution, finding ways to capture the sun's energy, transform it, and return it eventually to the environment through a chain of trophic exchanges. Thus, functional redundancy arises naturally. These features reduce the potential for catastrophic avalanches of extinction, to some extent restricting such cascades to the individual clusters and providing the substitute species to replace those displaced. Thus, a natural form of resiliency emerges from the evolution of the system's members, not from selection operating at the level of the whole system. The same trends are evident in any free-market economy, as Adam Smith argued.

The preceding paragraphs lay out two powerful and contrasting theories for the evolution and organization of ecosystems. In one, self-organized criticality, systems assemble themselves into an attracting state that is robust to external influence in its most superficial features but highly inconstant in terms of its membership. According to the other theory, the evolutionary view, variation and selection operating at levels below that of the ecosystem lead to modularity, heterogeneity, and diversity, perforce providing a system resiliency that emerges from the buffering of individual species and groups of species to change. Neither view is entirely correct; neither is entirely wrong. The exciting challenge in the study of complex adaptive systems in general, and ecosystems in particular, is to determine how these forces interact across a range of scales of space and time.

9

WHERE DO WE GO FROM HERE?

Complexity and the Commons

Over hundreds of millions of years of change, evolution has infused complexity and diversity into the design of organisms and the structure of the web of life. As complexity emerged, so, too, did the multiplicity of levels at which evolution could act, thereby creating and shaping individual behaviors, ecological communities, and human societies. Ecosystems have evolved some degree of resiliency in the face of environmental change, but what resiliency exists has emerged from selection acting on the components of the system, not from forces acting at the level of the entire ecosystem. There are no guarantees, and the potential for disaster is real, as exemplified by the desertification of previously productive areas, global pandemics of disease, and the collapse of marine fisheries.

Evolution operates most forcefully when feedback loops are tight. This occurs most naturally when individuals interact primarily with a small subset of the universe, realizing on realistic time scales the costs and benefits of their actions. In this way, coalitions and mutualisms form, building modularity that confers some resiliency to ecological systems. Evolution also builds diversity, reinforcing mutalism, recombination, and processes at higher levels of organization that allow genomic lines of descent to adapt to an ever-changing environment. Modularity itself helps maintain heterogeneity by building barriers to invasion, thereby sustaining local assemblages and mutualism. There are, in the nature and history of the evolutionary process, deep lessons for how we must manage our natural resources—maintaining diversity and modularity and recognizing that we are part of a complex adaptive system called the biosphere. We each operate daily in ways that are most strongly influenced by our own perceived benefits. We are more likely to keep our houses free of our trash than our communities, more likely to maintain our community's environment than our nation's, more likely to show concern for our nation's environment than the globe's. We are also more likely to worry about what things look like now than what they will look like in ten years, much less one hundred years, and the future beyond that is so distant as to be hardly a factor in influencing our behavior. In economic parlance, we heavily discount the future. Thus, in both space and time, the immediate takes precedence over the distant. We live in a global commons but exhibit patterns of behavior that reflect, in Garrett Hardin's term, the Tragedy of the Commons.[1] Recycling programs work to some extent because they make us feel good about ourselves, and to some extent because they are associated with rewards and penalties. Few individuals act in the way they do because they are convinced that their individual behaviors make much difference; they usually don't. But collective actions may make a great deal of difference. People vote in elections; if no one did, democracies would be in a great deal of trouble in choosing leaders. So it is possible, at least in theory, to convince people that individual actions are worth taking in support of communal governance.

The implications of this are that it is relatively easy to get neighborhood associations together to worry about zoning and shopping malls and the decay of the local environment, and that it is possible to convince people of the importance of having laws that restrict local pollution. As problems become broader in scale, however, the feedback loops become looser, the motivation becomes less, and the challenges for environmental action become greater. Achieving sustainability requires us to preserve the same options for future generations that are available to us now, but it is human nature to discount the importance of the future in relation to our needs and pleasures today. Similarly, environmental protection has been somewhat successful in dealing with local and regional pollution, but much less so when the threats are more global and hence the rewards for restraint are spread out over many people. The payback to any individual or corporation or nation lessens when the scale broadens. Hence, the incentives for cooperation are reduced, and the temptations to cheat on any agreement are increased.

Evidence of this phenomenon is to be found in the relationship of environmental quality to the wealth of nations. Gene Grossman and Alan Krueger, economists at Princeton, have shown that if one compares nations according to their per capita incomes, a rather complicated relationship emerges for many measures of environmental quality: with increasing income, there is more environmental degradation up to a point, after which environmental quality improves as income increases.[2] This relationship applies to a variety of measures, including sanitation, water quality, suspended particulates, carbon monoxide, sulfur dioxide, and nitrous oxides, and it has a variety of explanations.[3] The increasing degradation that one sees with increasing income at low income levels is undoubtedly associated simply with increased industrialization, but the environmental improvement with higher income levels is due to more subtle factors. Wealthier nations may indeed be in a better position to turn their attention to environmental quality. That may lead to such measures as developing less polluting alternatives to conventional practice, but it may also lead to the export of pollution, for example, by establishing factories in other nations or simply by purchasing goods that have been produced elsewhere. It is easy for us to be concerned about pollution from industrial sources if the pollution occurs in our own backyards; it is more difficult if we do not have to deal personally with the consequences of the pollution we support.

In the days since Rachel Carson alerted us to the problems of a decaying environment, we have come a long way in dealing with its degradation.[4] Laws have been enacted, and occasionally enforced, to deal with the most obvious forms of deterioration, especially air and water pollution in our local environments. These are situations where the signal is very clear, and where cause and effect are often easily linked. In such cases, we can detect the problem early, diagnose the solution, and introduce legislation to get straight to the source. That leaves a class of problems, however, that are much more refractory. Global climate change and the loss of biodiversity, for example, represent threats of a magnitude as great as any that have attracted public attention, but their effects take much longer to detect, and the causes are multiple and much less easily held culpable. It takes time to recognize the signal in the midst of noise, and the risk exists that irreversible changes will occur before actions can be taken. In the case of local pollution or neighborhood development, the homeostatic mechanisms are more effective: evidence of decay is quickly recognized and spurs actions that may correct the situation.

This is the Tragedy of the Commons. Environmental groups urge us to "think globally, and act locally," but until people also think locally, their motivation to act responsibly is weak. Making the payoffs for behavior in the common good nearer and clearer increases the chances of success. Sound and responsible environmental management demands equitable and *sustainable stewardship* of common resources. It requires recognizing that the biosphere is a highly complex adaptive system and learning how to harmonize human activities with the rest of that system. Through this program of action, we can harness the forces of evolution and self-organization for the common good. To do otherwise would be both to miss an opportunity and to run counter to natural forces of irresistible power.[5]

The Eight Commandments of Environmental Management

Given the biblical references earlier, I planned to end this treatise with ten commandments of environmental management. At the end of the day, however, I find that things are simpler than I thought. I did not need ten; the eight that follow summarize the essential management lessons emerging from our examination of the development and evolution of the biosphere. Even these overlap and intergrade with one another, but they provide a framework for sound practice.

1. Reduce uncertainty. This commandment may perhaps seem so obvious that it barely deserves mention. Yet our shocking ignorance of many of the most fundamental features of the world around us is the result in large part of an inadequate global commitment to acquiring the essential knowledge. Strangely, we seem willing to spend more money to search for life on other planets than to study diversity on our own. Not only are the vast majority of species still unknown scientifically, but we also do not have a very good fix on how many species there are. Knowledgeable estimates range from a few million to as many as 100 million,[6] with the general consensus between 10 and 20 million.[7] New species are being discovered all the time, from the tropical rain forests to the hydrothermal depths of the oceans, but species also are disappearing far faster than they can be identified. Furthermore, knowing what is present is only a first step in understanding how ecosystems function and in elucidating the importance of biological diversity to that functioning.

There is more than one way to reduce uncertainty. Monitoring and scientific investigations, of course, provide the core. Information so garnered informs policy debates, especially when it is shared with the public and its representatives. Another strategy, however, is to spread risks by broadening the scales at which we rely on the ecosystem for services. We have seen how evolution fosters the development of a diversity of mechanisms—including, for example, dispersal and dormancy, perenniality and iteroparity, plasticity and omnivory—by which individual genotypes may average over a range of environmental conditions. We also know that investment advisers urge their clients to diversify their stock portfolios; indeed, those who have the least confidence in their ability to predict or the least tolerance for variation should diversify the most. So, too, must the world's people reduce the effects of uncertainty by minimizing reliance on particular sources of energy or particular stocks of fish. The ability to do so is constrained by economic substitutability and fungibility—the potential, for example, to encourage consumers to shift from scarce natural resources to plentiful ones. There is, nonetheless, considerable scope for diversification, which is an imperative for survival.

The formation of family groups, communities, and societies is another device for protecting ourselves against environmental vagaries; such groupings increase the flow of information and provide insurance against individual misfortunes. Trade has the potential to extend those benefits within and among regions, from the cattle markets of India[8] to the global community of nations.[9] Trade, for example, allows the inter-regional averaging of years of famine and plenty, transferring excess production in some areas to others where it is more needed.

2. Expect surprise. I borrow here a page from the book of Crawford "Buzz" Holling, the University of Florida scientist who has done more than any other ecologist to help develop the science of adaptive management.[10] Adaptive management is maintaining flexibility in management structures and adjusting rules and regimes on the basis of monitoring and other sources of new data. It puts a premium, of course, on information, not just about the current status of a system but also about the likely consequences of alternative strategies for management or exploitation of a resource. Animals face such problems regularly, for example, in deciding the value of continuing to forage in a particular habitat. When a good habitat is found, there obviously is merit in sticking with it. Fidelity to a productive site provides a reliable source of nourishment; it also provides excellent information about local quality. The flip side of the coin, however, is that staying for too long in one place, or with one strategy, reduces knowledge about what is going on elsewhere. This is a serious problem: since competitive advantage translates into evolutionary success, there is a continual need for organisms to acquire information about the world beyond. Hence, the strategy of staying put can be improved by at least occasional exploratory forays to learn about conditions in other places. Fishermen, who must decide when to explore new fishing grounds, are prototypical examples of animals that face this trade-off, but we each confront similar issues in all aspects of our daily lives. In the case of environmental management, this necessity argues for *adaptive probing*—that is, for some continual exploration of alternative management strategies, even when current strategies seem to be working adequately.

Surprise is nothing more than the manifestation of the ineluctable core of uncertainty that Ralph Gomory has identified as key to distinguishing the *knowable* from the *unknowable*.[11] There is much that we know, and much that we do not know. The first commandment of environmental management argues that we should transfer as much as we can from the latter category to the former; the second commandment requires that we recognize that there are limits to knowledge and predictability, and that management must be appropriate to the essential element of uncertainty. Sometimes that may mean building flexible response systems so that the effects of unpredictable changes are reduced in their impact; other times it may mean finding mechanisms to build in resiliency to reduce the likelihood of qualitative shifts in services. Ultimately, of course, there may be unpredictable changes of such magnitude that no amount of preparation can help a great deal. Coupled models of the ocean and atmosphere, for example, suggest that the climate system could flip dramatically, in a period of a few years or less, between a relatively warm climate and a glacial one.[12] Wallace Broecker and his colleagues suggested in 1985 that such a flip could account for the spectacularly rapid transition from the warmer Allerod period to the glacial Younger Dryas 11,000 years ago.[13] The potential for such dramatic shifts argues above all for minimizing human activities that might engender them, as well as for taking whatever limited measures may be feasible to buffer humanity against the potentially catastrophic consequences.

Adaptive management is easier said than done. Fisheries have utilized such schemes for decades—setting quotas based on present information and revising them on the basis of new facts. But it is not always the case, even in fisheries, that the needed information is acquired rapidly enough to be useful. Indeed, the delays inherent in implementing new strategies based on old information can potentially destabilize a regulatory scheme. In the case of global change and biodiversity loss, this problem is exaggerated, because the time scale on which information is accumulating about the loss of ecosystem services is longer than the scale needed to implement changes. In situations such as this, there is therefore a strong argument for the invocation of precautionary principles where possible, even though this may involve economic costs that engender vigorous opposition.[14] An example might be the establishment of refuge areas, where fishing or hunting is not allowed. More generally, it means erring on the side of underexploitation of resources.

3. Maintain heterogeneity. Natural selection acts on the variability in a population; without heterogeneity, there is nothing for it to do. The resiliency of any complex adaptive system is embodied in its diversity and in the capacity for adaptive change among system components. Multiple crops and multiple varieties in agriculture, for example, have been introduced as devices to guard against massive failures and disease outbreaks, which are far more likely to occur in homogeneous environments.

The capability of any ecological system to respond to new environmental challenges is to be found in the maintenance of biological diversity, in all its forms. As we lose species, we lose the diversity and resiliency of ecosystems. As we lose tropical rain forests, wetlands, and other habitats, we lose not just species but the diversity of ecosystem types that sustains the functioning of the biosphere.

Natural ecosystems are dynamic arenas in which variability and change at local scales maintain diversity at broader ones. The vast majority of plant species are locally ephemeral and rely for their survival on the transient windows of opportunity provided by frequent and stochastic local disturbances, such as those caused by fire, weather, waves, and windthrows. Thus, management efforts to reduce the effects of such influences weaken the capacity of the system to respond. If there is balance in nature, it is to be found only at the broadest scales of space, time, and organizational complexity.

4. Sustain modularity. The parable of the two watchmakers illustrates a point that holds deep lessons for management, whether of corporations or of the biosphere: in modular structures, there is buffering against cascades of disaster. Per Bak argues that complex systems naturally achieve self-organized criticality.[15] As we saw in the last chapter, however, complex adaptive systems witness evolutionary forces toward modularization and compartmentalization that buffer systems against such cascades. Still, increasing trends toward globalization make the world a smaller place, and weaken modularization.[16] Especially dramatic effects may be seen in the interconnectedness of international financial markets and in the emergence and reemergence of global pandemics of disease. Diseases such as AIDS, for example, may well have been contained within small, insulated populations a few decades ago; today few populations remain isolated, and the interconnectedness of peoples facilitates rapid spread. Modularity has broken down, compromising and threatening the resiliency of humanity.

5. Preserve redundancy. The maintenance of heterogeneity and modularity, of course, helps maintain redundancy of function. Redundancy and heterogeneity are hand and glove; much redundancy is reflected, for example, in the heterogeneity within functional groups of species performing similar ecological roles. Redundancy, however, can also exist in the absence of heterogeneity: it provides the Ehrlichs' extra rivets, those system components whose value becomes clear only when other parts are lost. Like the rivets, it is their identity of function that characterizes species redundancy, not the individuality of the actors. Redundancy is the immediate source of replacement of lost functions; heterogeneity provides the material for adaptive responses over longer time scales. Furthermore, the heterogeneity within a functional group, as discussed earlier, reduces the functional susceptibility to specific threats. The essential element to understanding the importance of redundancy is to elucidate the functional substitutability of one species for another, the ecological complement to economic substitutability. Current research on the relationship between biodiversity and ecosystem function is rapidly illuminating this central issue, but we have much to learn.

6. Tighten feedback loops. One of the central lessons emerging from our examination of the evolution of interactions between organisms is that tight reward and punishment loops are essential for any adaptive change. Unfortunately, those feedback loops are weak or absent in regard to the utilization of common resources, as well as for problems of global extent. Where many individuals exploit the same resource, or contribute to the same problem, they may not see their own individual actions as relevant; more important, given that evolution shapes organisms to operate selfishly, or at best in the self-interest of their genes, the benefits to individuals in restraining their own consumption or exploitation patterns are faint in the case of common resources.

To encourage behavior in the common good, we must tighten cost and benefit feedback loops. Economists urge reliance on the market to solve problems, but the market can work properly only when pricing reflects the true value of a good, including the social costs. Fish populations belong to the world at large; there is thus little motivation for an individual fisherman to limit catch. An uncaught fish is a lost opportunity. The mobile nature of fish populations makes ownership a difficult concept, but it is worth comparing the situation to the preservation of fugacious species in Africa, where the transfer of certain property rights in charismatic endangered species to local human populations has fostered conservation.[17] *Privatization* and the development of ecotourism have given a value to these animals that tightens feedback loops and provides people with a motive for preservation.[18] This does not mean that privatization is without its risks. Assigning property rights to some individuals may thereby deny them to others, in extreme cases creating destitutes[19] as well as fomenting rebellion.[20] Moreover, privatization does not guarantee conservation if economic discount rates are such that it pays to overexploit now rather than to preserve for the future.[21] On balance, however, the potential benefits are great if the system is devised wisely and fairly.

Graciela Chichilnisky and Geoffrey Heal, two economists at Columbia University, have powerfully demonstrated through economic arguments how privatization can help the markets to "get the price right" and thereby encourage conservation measures.[22] Heal highlights the work of the Conservation Corporation, based in South Africa, which has returned huge areas in Africa to their original ecological status in order to increase their value for ecotourism and hunting.[23] Land that was worth less than $100 per hectare when used for agriculture can increase in value by as much as an order of magnitude, providing strong incentive for protection. Chichilnisky and Heal, in fundamentally important work, have also demonstrated how New York City realized billions of dollars in economic benefits by sustaining the Catskills watershed as a water filtration system, rather than destroying it completely and building a new filtration plant.[24]

Tightening feedback loops leads to empowerment, giving people incentives for environmentally beneficial behavior. The return to local control of the Mudialy wetlands near Calcutta not only improved fishery productivity but also provided a harvest of wood.[25] Similarly, overpopulation is a problem primarily in areas where women have no power and few options but to stay home and produce children to provide the "extra hands" needed to help with the work of the family. With education and liberalization of social systems comes empowerment; in more developed nations, women are forces in the workplace, and population growth is a disappearing problem.

Tightening feedback loops also means getting the prices right. To find ways to charge properly for the exploitation of common resources, we must make intelligent use of appropriate taxes on pollution, including, for example, the use of carbon taxes to reduce harmful releases into the environment. Air and water, like fish populations, are common property resources whose use as receptacles for effluents is never adequately paid for by those who release materials into them. Finally, perverse subsidies that harm the environment must be eliminated and replaced by reward schemes that encourage good environmental practices. These include reforestation, which can have dramatic effects on restoring balance to the carbon cycle, as well as exploration and innovation in technologies that conserve energy and reduce pollution.

7. Build Trust. Global problems require global solutions, but much can be learned from successful strategies for dealing with local and regional problems. As we have seen throughout this book, evolution works most effectively when individuals interact most with their near neighbors. Through the experiences gained by repeated interactions, trust can develop, and strategies such as reciprocal altruism may be favored. This can be extended to regional cooperatives, such as that involving 15,000 farmers sharing the waters of the River Tieria, near Valencia, Spain. A local tribunal, which has met on Thursday mornings for centuries to hear complaints about cheating, has kept this system operational for 550 years.[26] The key is that the community is still small enough for trust to be maintained.[27] The River Tieria tribunal occupies a middle ground between complete privatization and complete centralization.[28] At the global level, such trust is missing; through centuries of conflict and competition, the same forces that build trust within cultures pit one against another.

Yet treaties exist, and nations by and large meet their obligations.[29] Trust develops, reinforced by the weight of the community of nations. As we know from many examples throughout history, the control mechanisms are weak and do not effectively restrain renegade nations. Yet they can work, however tenuously.

International agreements have had mixed success in achieving environmental goals. Among the most successful has been the Montreal Protocol, the accord that is the basis for the removal of chlorofluorocarbons (CFCs) and other ozone-depleting substances from the environment.[30] The Montreal Protocol has involved nearly every country in the world and was brilliantly designed to be robust in the face of occasional cheaters. There is a stable body of trust, increasing the likelihood that the accord will achieve its central goals. The protocol is a model example of environmental cooperation; unfortunately, recent efforts to imitate it in the areas of global change and biodiversity have not enjoyed comparable success. Furthermore, years of international efforts to regulate whaling and other fisheries have also provided a mixed bag: the nations have failed to develop the necessary trust, and some have been unwilling to suppress their own interests for the greater good.

One can cajole and censure recalcitrant states, but world opinion is in general an ineffective mechanism to get nations to do what they don't want to do. Where one nation can influence another's decisions, it is in general because of enforceable or implied compacts, most generally involving a range of issues. These are, in essence, what maintains trust. There are other mechanisms, however, until now poorly exploited. Developing nations see their own resources, such as rain forests, in a different light than do the more developed countries. For indigenous peoples, land has value in terms of meeting the immediate needs of an expanding population; for others, it provides insurance against global problems much further down the road. Different cultures discount the future at different rates, and various theories exist for how they should do so. At the core of resolving conflicting views, however, must be a balancing of intergenerational equity and intragenerational equity.[31] If we in the developed world have an interest in encouraging developing nations to conserve their vanishing resources toward a sustainable future, then we must not only help them to meet their immediate needs in other ways but also provide them with incentives—such as aiding the development of ecotourism, or the exchange of their debts for nature preservation—to tighten the reward loop and make conservation practice a benefit for the local and national decisionmakers.[32] More generally, wherever the interests of different nation-states are involved, we must find ways for all people and nations to realize their own self-interest in the common good.[33]

8. Do Unto Others as You Would Have Them Do Unto You. The great rabbi Hillel, when taunted by a skeptic to expound all of the principles of Jewish law while he perched on one foot, said, "Whatever is hateful unto thee, do it not unto thy fellow: this is the whole Torah; the rest is explanation."[34] This, the so-called Golden Rule, can be found in one form or another in the teachings of most religions. Its essential good sense provides the foundation on which societies are built, independent of the role of religion in those societies. Societies can survive only when there is action for the collective good, whether maintained by reciprocal altruism, binding agreements, systems of law, or international compacts. Building trust and environmental security is not an easy task, but it provides the only path to sustaining our fragile dominion over the Earth's resources.

NOTES

Chapter One

1. Primeval light existed on day one and apparently sufficed for some photosynthesis, though the sun did not make its appearance until day four. Joseph H. Hertz, *The Pentateuch and Haftorahs,* 2nd ed. (London: Soncino Press, 1963), 2–4.

2. Hertz, *The Pentateuch and Haftorahs,* 27.

3. Paul R. Ehrlich and Simon A. Levin, "Biodiversity: What It Is and Why We Need It," in Linda Koebner, Jane E. S. Sokolow, Francesco T. Grifo, and Sharon Simpson, eds., *Scientists on Biodiversity* (New York: American Museum of Natural History, 1998), 20–23; Jennifer Hughes, Gretchen C. Daily, and Paul R. Ehrlich, "Population Diversity: Its Extent and Evolution," *Science* 278 (1997): 689–692.

4. Ian Turner, "Singapore: A Case Study for Tropical Rainforest Fragmentation and Biodiversity Loss," in Takuya Abe, Simon A. Levin, and Masahiko Higashi, eds., *Biodiversity: An Ecological Perspective* (New York: Springer, 1997), 249–258.

5. Gretchen C. Daily, ed., *Nature's Services: Societal Dependence on Natural Ecosystems* (Washington, D.C.: Island Press, 1997).

6. Margery L. Oldfield, *The Value of Conserving Genetic Resources* (Washington, D.C.: U.S. Department of the Interior, National Parks Service, 1984).

7. Ibid.

8. Sandra Postel and Stephen Carpenter, "Freshwater Ecosystems Services," in Daily, *Nature's Services,* 195–214.

9. L. R. Oldeman, V. van Engelen, and J. Pulles, "The Extent of Human-Induced Soil Degradation, Annex 5," in L. R. Oldeman, R.T.A. Hakkeling, and Wim G. Sombroek, eds., *World Map of the Status of Human-Induced Soil Degradation: An Explanatory Note* (Wageningen: International Reference and

Information Center, 1990); D. Pimentel, C. Harvey, et al., "Environmental and Economic Costs of Soil Erosion and Conservation Benefits," *Science* 267 (1995): 1117–1123.

10. Norman Myers, "The World's Forests and Their Ecosystem Services," in Daily, *Nature's Services,* 215–235.

11. Ibid.; W. Magrath and P. Arens, *The Costs of Soil Erosion on Java: A Natural Resource Accounting Approach* (Washington, D.C.: World Bank, 1989).

12. Aldo Leopold, *Round River* (New York: Ballantyne Books, 1970), 190.

13. Paul R. Ehrlich and Anne H. Ehrlich, *Extinction: The Causes and Consequences of the Disappearance of Species* (New York: Random House, 1981), xii–xiii.

14. Murray Gell-Mann, *The Quark and the Jaguar: Adventures in the Simple and the Complex* (New York: W. H. Freeman and Co., 1994).

15. Simon A. Levin, "Ecosystems and the Biosphere as Complex Adaptive Systems," *Ecosystems* 1 (1995): 431–436.

16. Gell-Mann, *The Quark and the Jaguar,* 16.

17. John Holland, *Hidden Order: How Adaptation Builds Complexity* (Reading, Mass.: Addison Wesley, 1995).

18. Ibid., 10–12.

19. W. Brian Arthur, *Increasing Returns and Path Dependence in the Economy* (Ann Arbor: University of Michigan Press, 1994).

Chapter Two

1. Lawrence B. Slobodkin, "The Strategy of Evolution," *American Scientist* 52 (1964): 342–357.

2. François Jacob, "Evolution and Tinkering," *Science* 196 (1977): 1161–1166.

3. James E. Lovelock, and Lynn Margulis, "Atmospheric Homeostasis by and for the Biosphere: The Gaia Hypothesis," *Tellus* 26 (1974): 2–9.

4. David Abram, "The Mechanical and the Organic: On the Impact of Metaphor in Science," in Stephen H. Schneider and Penelope J. Boston, eds., *Scientists on Gaia* (Cambridge, Mass.: MIT Press, 1991), 66–76.

5. James E. Lovelock, "Gaia as Seen Through the Atmosphere," *Atmospheric Environment* 6 (1972): 579–580.

6. Paul Ehrlich, "Coevolution and Its Applicability to the Gaia Hypothesis," in Schneider and Boston, *Scientists on Gaia* 19–22.

7. Ibid.

8. Schneider and Boston, Preface to *Scientists on Gaia,* xiii–xv.

9. Lynn Margulis and Gregory Hinkle, "The Biota and Gaia: 150 Years of Support for Environmental Science," in Schneider and Boston, *Scientists on Gaia*, 11–18.

10. William J. Burroughs, *Weather Cycles: Real or Imaginary* (Cambridge: Cambridge University Press, 1992).

11. Lovelock and Margulis, "Atmospheric Homeostasis by and for the Biosphere."

12. Ehrlich, "Coevolution and Its Applicability to the Gaia Hypothesis."

13. Richard C. Lewontin, "Adaptation," *Enciclopedia Einaudi Turin* 1 (1977): 198–214.

14. Paul Davies, *The Runaway Universe* (New York: Harper & Row, 1978), 185.

15. Robert B. Hargraves, "Precambrian Geologic History," *Science* 193 (1976): 363–371; J. Graham Cogley and Ann Henderson-Sellers, "The Origin and Earliest State of the Earth's Hydrosphere," *Reviews of Geophysics* 22 (1984): 131–175.

16. Brian Henderson-Sellers, Ann Henderson-Sellers, S. Mary P. Benbow, and Kendal McGuffie, "Earth—The Water Planet: A Lucky Coincidence?" in Schneider and Boston, *Scientists on Gaia*, 80–89.

17. Ehrlich, "Coevolution and Its Applicability to the Gaia Hypothesis"; Timothy M. Lenton, "Gaia and Natural Selection," *Nature* 394 (1998): 439–447.

18. James E. Lovelock, "Gaia as Seen Through the Atmosphere," in Peter Westbroek and Elizabeth W. deJong, eds., *Biomineralization and Biological Metal Accumulation* (Dordrecht: Reidel, 1983), 15–25; Andrew J. Watson and James E. Lovelock, "Biological Homeostasis of the Global Environment: The Parable of Daisyworld," *Tellus* 35B (1983): 284–289.

19. Lenton, "Gaia and Natural Selection."

20. Luigi L. Cavalli-Sforza and Marcus W. Feldman, *Cultural Transmission and Evolution: A Quantitative Approach* (Princeton, N.J.: Princeton University Press, 1981).

Chapter Three

1. Stephen Jay Gould, *Wonderful Life* (New York: Norton, 1989), 45–52.

2. F. Herbert Bormann and G. Eugene Likens, *Pattern and Process in a Forested Ecosystem* (New York: Springer, 1979).

3. Stephen R. Carpenter and James F. Kitchell, eds., *The Trophic Cascade in Lakes* (Cambridge: Cambridge University Press, 1993); Robert T. Paine, *Marine Rocky Shores and Community Ecology: An Experimentalist's Perspective* (Oldendorf/Luhe, Germany: Ecology Institute, 1994); David Tilman,

Plant Strategies and the Dynamics and Structure of Plant Communities, Monographs in Population Biology, vol. 26 (Princeton, N.J.: Princeton University Press, 1988); David Tilman and John A. Downing, "Biodiversity and Stability in Grasslands," *Nature* 367 (1994): 363–365.

4. Per Bak and Kan Chen, "Self-organized Criticality," *Scientific American* 2641 (1995): 46–54.

5. For a power law distribution of earthquake intensity, the frequency of earthquakes of a given intensity is proportional to a negative power of intensity. Power laws are explored in more detail in later chapters.

6. Stuart Kauffman, *The Origins of Order* (New York: Oxford University Press, 1993). See also note 14, chapter 8.

7. Stuart Kauffman, *At Home in the Universe* (New York: Oxford University Press, 1995).

8. Per Bak, *How Nature Works* (New York: Springer-Verlag, 1996).

Chapter Four

1. François Jacob, *The Possible and the Actual* (New York: Pantheon, 1982).

2. Ray Hilborn and Marc Mangel, *The Ecological Detective: Confronting Models with Data* (Princeton, N.J.: Princeton University Press, 1997).

3. "Adam Smith" [George J. W. Goodman], *The Money Game* (New York: Vintage, 1976), 136.

4. W. Brian Arthur, "Positive Feedbacks in the Economy," *Scientific American* (February 1990): 90–92.

5. Paul David, "Clio and the Economics of QWERTY," *American Economic Review Proceedings* 75 (1986): 332–337.

6. Stuart Kauffman and Simon A. Levin, "Towards a General Theory of Adaptive Walks on Rugged Landscapes," *Journal of Theoretical Biology* 1281 (1987): 11–45.

7. Jacob, *The Possible and the Actual;* François Jacob, "Evolution and Tinkering," *Science* 196 (1977): 1161–1166.

8. William M. Wheeler, *Ants: Their Structure, Development, and Behavior* (New York: Columbia University Press, 1910).

9. Daniel H. Janzen, "Coevolution of Mutualism Between Ants and Acacias in Central America," *Evolution* 20 (1966): 249–275.

10. Jared Diamond and Ted J. Case, "Overview: Introductions, Extinctions, Exterminations and Invasions," in Jared Diamond and Ted J. Case, eds., *Community Ecology* (New York: Harper & Row, 1986), 65–79.

11. I.A.E. Atkinson, "Effects of Rodents on Islands," in P. J. Moors, ed., *Conservation of Island Land Birds* (Cambridge: International Council for Bird Preservation, 1985), 35–81.

12. Diamond and Case, "Overview."

13. Jacob, "Evolution and Tinkering."

14. Lawrence B. Slobodkin, "The Strategy of Evolution," *American Scientist* 52 (1964): 342–357.

15. Leslie R. Holdridge, "Determination of World Plant Formations from Simple Climatic Data," *Science* 105 (1947): 367–368.

16. Stuart L. Pimm, *Food Webs* (New York: Chapman & Hall, 1982).

17. Joel E. Cohen, *Food Webs and Niche Space*, Monographs in Population Biology, vol. 11 (Princeton, N.J.: Princeton University Press, 1978).

18. Gary A. Polis, "Complex Trophic Interaction in Deserts: An Empirical Critique of Food-web Theory," *American Naturalist* 138 (1991): 123–155.

19. Gordon H. Orians and Otto T. Solbrig, *Convergent Evolution in Warm Deserts* (Stroudsburg, Penn.: Dowden, Hutchinson and Ross, 1977).

20. Andreas F. W. Schimper, *Plant Geography upon a Physiological Basis* (Oxford: Oxford University Press/Clarendon Press, 1903).

21. See, for example, Martin Cody, "Structural Niches in Plant Communities," in Diamond and Case, *Community Ecology*, 381–405.

Chapter Five

1. Sherwin Carlquist, *Island Life: A Natural History of Islands of the World* (Garden City, N.Y.: Natural History Press, 1965).

2. Robert H. MacArthur and Edward O. Wilson, *The Theory of Island Biogeography*, Monographs in Population Biology 1, vol. 1 (Princeton, N.J.: Princeton University Press, 1967).

3. Daniel Simberloff and Edward O. Wilson, "Experimental Zoogeography of Islands: The Colonization of Empty Islands," *Ecology* 50 (1969): 278–296.

4. Willem M. Doctors van Leeuwen, "Krakatau's New Flora," *Proceedings of the Fourth Pacific Science Congress* (Batavia), part 2 (1929): 56–61.

5. Colin S. Reynolds, "Successional Development, Energetics, and Diversity in Planktonic Communities," in Takuya Abe, Simon A. Levin, and Masahiko Higashi, eds., *Biodiversity: An Ecological Perspective* (New York: Springer, 1997), 167–202.

6. Martin S. Nowak, Robert M. May, and Roy M. Anderson, "The Evolutionary Dynamics of HIV Quasispecies and the Development of Immunodeficiency Disease," *AIDS* 4 (1990): 1095–1103; Stephen M. Wolinsky et al., "Adaptive Evolution of Human Immunodeficiency Virus–Type 1 During the Natural Course of Infection," *Science* 272 (1996): 537–542; Simon A. Levin, Bryan Grenfell, Alan Hastings, and Alan S. Perelson, "Mathemati-

cal and Computational Challenges in Population Biology and Ecosystem Science," *Science* 275 (1997): 334–343.

7. Roland H. Lamberson, Barry R. Noon, Curt Voss, and Kevin S. McKelvey, "Reserve Design for Territorial Species: The Effects of Patch Size and Spacing on the Viability of the Northern Spotted Owl," *Conservation Biology* 8 (1994): 185–195; Barry R. Noon and Kevin S. McKelvey, "Management of the Spotted Owl: A Case Study in Conservation Biology," *Annual Review of Ecology and Systematics* 27 (1996): 135–162.

8. Simon A. Levin and Stephen W. Pacala, "Theories of Simplification and Scaling of Spatially Distributed Processes," in David Tilman and Peter Kareiva, eds., *Spatial Ecology: The Role of Space in Population Dynamics and Interspecific Interactions* (Princeton, N.J.: Princeton University Press, 1997), 271–296; Stephen W. Pacala and Simon A. Levin, "Biologically Generated Spatial Pattern and the Coexistence of Competing Species," in Tilman and Kareiva, *Spatial Ecology,* 204–232; Benjamin M. Bolker, Stephen W. Pacala, and Simon A. Levin, "Moment Methods for Stochastic Processes in Continuous Space and Time," in Ulf Dieckmann and J.A.J. (Hans) Metz, eds., *Proceedings of "Low-Dimensional Dynamics of Spatial Ecological Systems,"* November 14–16, 1996, Laxenburg, Austria (in press).

9. The voter model and the contact process are variants on models that have played important roles in physics, allowing the elucidation of fundamental principles. The Ising model, which actually was first proposed by the physicist Herbert Lenz but bears the name of his student Ernst Ising, was introduced to explain and explore the properties of magnets. When a permanent magnet is heated, it loses magnetization. The Lenz-Ising model imagines tiny magnets on a regular lattice, each with a "spin" that is either up or down. A spin may flip from up to down, or vice versa, according to criteria that depend (thermodynamically) on the spins of neighbors. From these simple local rules of interaction, macroscopic or bulk phenomena—such as the loss of a magnet's properties as it is heated—emerge. Lenz-Ising materials are special cases of "spin-glasses," disordered magnetic materials made up of large numbers of interacting magnetic dipoles. Models of this sort have found explicit application in a variety of fields beyond physics—for example, the "N-K" models of Kauffman and Levin as they relate to the theory of coevolution (Stuart Kauffman and Simon A. Levin, "Towards a General Theory of Adaptive Walks on Rugged Landscapes," *Journal of Theoretical Biology* 1281 [1987]: 11–45).

10. Simon A. Levin, "The Problem of Pattern and Scale in Ecology," *Ecology* 73 (1992): 1943–1967.

11. Gordon H. Orians and Otto T. Solbrig, *Convergent Evolution in Warm Deserts* (Stroudsburg, Penn.: Dowden, Hutchinson and Ross, 1977).

12. Joel E. Cohen, *Food Webs and Niche Space,* Monographs in Population Biology, vol. 11 (Princeton, N.J.: Princeton University Press, 1978); Stuart L. Pimm, *Food Webs* (New York: Chapman & Hall, 1982); Gary A. Polis, "Complex Trophic Interaction in Deserts: An Empirical Critique of Food-Web Theory," *American Naturalist* 138 (1991): 123–155.

13. Pimm, *Food Webs*; Peter Yodzis, "The Structure of Assembled Communities," *Journal of Theoretical Biology* 92 (1981): 103–117; George Sugihara, *Niche Hierarchy: Structure, Organization, and Assembly in Natural Communities* (Ph.D. diss., Princeton University, 1982); James A. Drake, *Some Theoretical and Empirical Explorations of Structure in Food Webs* (Ph.D. diss., Purdue University, 1985).

14. David Tilman, David Wedin, and Johannes Knops, "Productivity and Sustainability Influenced by Biodiversity in Grassland Ecosystems," *Nature* 379 (1996): 718–720; David Tilman, Johannes Knops, David Wedin, Peter Reich, Mark Ritchie, and Evan Siemann, "The Influence of Functional Diversity and Composition on Ecosystem Processes," *Science* 277 (1997): 1300–1302.

15. National Research Council (NRC), Ocean Studies Board, *Sustaining Marine Fisheries* (Washington, D.C.: National Academy Press, 1999).

16. Food and Agriculture Organization (FAO), *Review of the State of World Fishery Resources: Marine Fisheries,* Circular 920 FIRM/C920 (Rome: FAO, 1997); NRC, *Sustaining Marine Fisheries;* Carl Safina, "The World's Imperiled Fish Stocks," *Scientific American* 273 (1995): 46–53.

17. Vito Volterra, "Variazioni e fluttuazioni del numero d'individui in specie animale conviventi," *Memorie della Reale Accademia Nazionale dei Lincei* (ser. 6) 2 (1926): 31–113, translated in Royal N. Chapman, *Animal Ecology* (New York: McGraw-Hill, 1931); Edward H. Kerner, "A Statistical Mechanics of Interacting Biological Species," *Bulletin of Mathematical Biophysics* 19 (1957): 121–146; Egbert G. Leigh Jr., "On the Relation Between the Productivity, Biomass, Diversity, and Stability of a Community," *Proceedings of the National Academy of Science* 53 (1965): 777–783; Egbert G. Leigh Jr., "The Ecological Role of Volterra's Equations," in Murray Gerstenhaber, ed., *Some Mathematical Problems in Biology* (Providence: American Mathematical Society 1968), 1–61.

18. Daniel Janzen, personal communication, April 1998.

19. W. Thomas Edmondson, "Lake Washington," in Charles R. Goldman, James McEvoy III, and Peter J. Richerson, eds., *Environmental Quality and Water Development* (San Francisco: W. H. Freeman and Co., 1973).

20. F. Herbert Bormann and G. Eugene Likens, *Pattern and Process in a Forested Ecosystem* (New York: Springer, 1979).

21. Henry H. Shugart, *A Theory of Forest Dynamics* (New York: Springer, 1984); Stephen W. Pacala, Charles D. Canham, and John A. Silander, "Forest Models Defined by Field Measurements: I. The Design of a Northeastern Forest Simulator," *Canadian Journal of Forest Research* 23 (1993): 1980–1988.

22. Pacala, Canham, and Silander, "Forest Models Defined by Field Measurements"; Benjamin M. Bolker, Stephen W. Pacala, Fakhri A. Bazzaz, Charles D. Canham, and Simon A. Levin, "Species Diversity and Ecosystem Response to Carbon Dioxide Fertilization: Conclusions from a Temperate Forest Model," *Global Change Biology* 1 (1995): 373–381.

23. Ralph E. Gomory, "The Known, the Unknown, and the Unknowable," *Scientific American* 272(6) (1995): 120.

24. John Holland, *Hidden Order: How Adaptation Builds Complexity* (Reading, Mass.: Addison Wesley, 1995).

25. Ibid.

26. Simon A. Levin and Robert T. Paine, "Disturbance, Patch Formation, and Community Structure," *Proceedings of the National Academy of Science* 71 (1974): 2744–2747; Robert T. Paine and Simon A. Levin, "Intertidal Landscapes: Disturbance and the Dynamics of Pattern," *Ecological Monographs* 51 (1981): 145–178.

27. Alexander S. Watt, "Pattern and Process in the Plant Community," *Journal of Ecology* 35 (1947): 1–22.

Chapter Six

1. See, for example, Jonathan Weiner's wonderful book *The Beak of the Finch: A Story of Evolution in Our Time* (New York: Knopf, 1994).

2. William B. Provine, *Sewall Wright and Evolutionary Biology* (Chicago: University of Chicago Press, 1986).

3. Sewall Wright, "The Roles of Mutation, Inbreeding, Crossbreeding, and Selection in Evolution," *Proceedings of the Sixth International Congress of Genetics* 1 (1932): 356–366.

4. François Jacob, "Evolution and Tinkering," *Science* 196 (1977): 1161–1166.

5. Stuart Kauffman and Simon A. Levin, "Towards a General Theory of Adaptive Walks on Rugged Landscapes," *Journal of Theoretical Biology* 1281 (1987): 11–45.

6. In this effort, we had a great deal of help from my student Danny Grünbaum, who did the programming and provided valuable feedback.

7. All of these techniques are directed toward finding the optima—that is, the largest or smallest values—of some function, hence the term "optimiza-

tion." This is one of the most important problems in applied mathematics, because of its wide applicability. Ways to minimize costs or maximize benefits from some process are obviously widespread in applications from business to the military. The optimization process, like the evolutionary process, involves search on a landscape. A (single-valued) *function* translates *independent variables,* such as the amount of a product produced, into a *dependent variable*, for example, profit. The independent variables, typically many, are spread over some space where the search takes place; the dependent variable assigns a value, essentially a fitness, to each point in that high-dimensional space. In principle, we may think of the dependent variable as defining a landscape floating above the space of independent variables. In practice, as with Wright's metaphor, it is impossible for us to visualize such a landscape easily if there are more than two independent variables. Many conventional methods require that one conduct the search by computing slopes on the landscape in various directions and (in the case of a maximization problem) moving in the direction of steepest slope. In a high-dimensional space, this can be computationally very expensive, because there are so many different directions. In contrast, so-called *direct methods* are popular because they do not involve computation of slopes but use other ways of guiding search. Direct methods include Dantzig's *simplex method,* as well as the closely related Nelder-Mead method, both of which are built on the use of geometric figures. Stochastic optimization, a very efficient direct approach, mimics evolution. It works by randomly sampling from the neighborhood of a point in the space of independent variables, moving to another point if, for example, a better value of the dependent variable is discovered. The modifier "stochastic" refers to the fact that neighbors are sampled randomly.

8. Shu Lin and Brian W. Kernighan, "An Effective Heuristic Algorithm for the Traveling-Salesman Problem," *Operations Research* 21 (1973): 498–516.

9. Provine, *Sewall Wright.*

10. In the theory of mathematical optimization, this is sometimes called *simulated annealing,* which borrows from concepts in statistical physics.

11. Sewell Wright, "Evolution in Mendelian Populations," *Genetics* 16 (1931): 97–159.

12. Provine, *Sewall Wright,* 283–285, 310–311, 484.

13. John Maynard Smith, "The Theory of Games and the Evolution of Animal Conflicts, *Journal of Theoretical Biology* 47 (1974): 209–221; John Maynard Smith, *Evolution and the Theory of Games* (Cambridge: Cambridge University Press, 1982).

14. H.B.D. Kettlewell, "The Phenomenon of Industrial Melanism in the Lepidoptera," *Annual Review of Entomology* 6 (1961): 245–262.

15. William L. Brown and Edward O. Wilson, "Character Displacement," *Systematic Zoology* 5 (1956): 49–64.

16. David Lack, *Darwin's Finches* (Cambridge: Cambridge University Press, 1947).

17. B. Rosemary Grant and Peter R. Grant, *Evolutionary Dynamics of a Natural Population: The Large Cactus Finch of the Galápagos* (Chicago: University of Chicago Press, 1989).

18. Dolph Schluter, Trevor D. Price, and Peter R. Grant, "Ecological Character Displacement in Darwin's Finches," *Science* 227 (1985): 1056–1059.

19. Joseph H. Connell, "Diversity and the Coevolution of Competitors, or the Ghost of Competition Past," *Oikos* 35 (1980): 131–138.

20. Richard C. Lewontin, "Adaptation," *Enciclopedia Einaudi Turin* 1 (1977): 198–214.

21. Edward O. Wilson, "The Species Equilibrium," in Brookhaven National Laboratory Biology Department, ed., *Diversity and Stability in Ecological Systems,* vol. 22 (Upton, N.Y.: BNL Biology Department, 1969), 38–47.

22. Robert E. Ricklefs, *Ecology* (New York: Chiron Press, 1973).

23. Mark Newman, personal communication, 1998.

24. Ann P. Kinzig and John Harte, "Selection of Microorganisms in a Spatially Heterogeneous Environment, and Implications for Plant Access to Nitrogen," *Journal of Ecology* (1998): 841–853.

25. Garrett Hardin, "The Tragedy of the Commons," *Science* 162 (1968): 1243–1248.

26. Samuel Butler, *Life and Habit* (London: Trübner Co., 1878).

27. William D. Hamilton, "The Genetical Evolution of Social Behavior," *Journal of Theoretical Biology* 7 (1964): 1–52.

28. Richard Dawkins, *The Selfish Gene* (New York: Oxford University Press, 1976).

29. Robert Axelrod, *Evolution of Cooperation* (New York: Basic Books, 1984).

30. John H. Miller, "The Coevolution of Automata in the Repeated Prisoner's Dilemma," *Journal of Economic Behavior* 29 (1996): 87–112.

31. Martin A. Nowak and Robert M. May, "Evolutionary Games and Spatial Chaos," *Nature* 359 (1992): 826–829; Richard T. Durrett and Simon A. Levin, "The Importance of Being Discrete and (Spatial)," *Theoretical Population Biology* 46 (1994): 363–394.

32. Durrett and Levin, "The Importance of Being Discrete and (Spatial)."

33. John Tyler Bonner, *Life Cycles* (Princeton, N.J.: Princeton University Press, 1993).

34. Ibid., 3.

35. See, for example, Stephen J. Gould and Richard C. Lewontin, "The Spandrels of San Marco and the Panglossian Paradigm: A Critique of the Adaptationist Programme," *Proceedings of the Royal Society of London* (ser. B 205) (1979): 581–598.

36. Calaway H. Dodson, Robert L. Dressler, Harold G. Hills, Ralph M. Adams, and Norris H. Williams, "Biologically Active Compounds in Orchid Fragrances," *Science* 164 (1969): 1243–1249.

37. Robert L. Dressler, "Pollination by Euglossine Bees," *Evolution* 22 (1968): 202–210.

38. Bruce R. Levin and Richard E. Lenski, "Coevolution in Bacteria and Their Viruses and Plasmids," in Douglas J. Futuyma and Montgomery Slatkin, eds., *Coevolution* (Sunderland, Mass.: Sinauer Associates, 1983), 99–127.

39. Lin Chao, *The Population of Colicinogenic Bacteria: A Model for the Evolution of Allelopathy* (Ph.D. diss., University of Massachusetts, 1979); Lin Chao and Bruce R. Levin, "Structured Habitats and the Evolution of Anticompetitor Toxins in Bacteria," *Proceedings of the National Academy of Science* 78 (1981): 6324–6328.

40. C. H. Muller, "The Role of Chemical Inhibition (Allelopathy) in Vegetational Composition," *Bulletin of the Torrey Botanical Club* 93 (1966): 332–351.

41. Paul R. Ehrlich and Peter H. Raven, "Butterflies and Plants: A Study in Coevolution," *Evolution* 18 (1964): 586–608.

42. David Sloan Wilson, *The Natural Selection of Populations and Communities* (Menlo Park, Calif.: Benjamin/Cummings, 1980).

43. W. Eichler, "Some Rules in Ectoparasitism," *Annals and Magazine of Natural History* 12 (1948): 588–598; Daniel R. Brooks and Charles Mitter, "Phylogenetics and Coevolution," in Quentin Wheeler and Meredith Blackwell, eds., *Fungus/Insect Relationships: Perspectives in Ecology and Evolution* (New York: Columbia University Press, 1983).

44. Douglas J. Futuyma, "Evolutionary Interactions Among Herbivorous Insects and Plants," in Futuyma and Slatkin, *Coevolution,* 207–231.

45. Tibor Jermy, "Insect-Host-Plant Relationship—Coevolution or Sequential Evolution?" *Symposia Biologica Hungarica* 16 (1976): 109–113.

46. Lawrence E. Gilbert, "Butterfly-Plant Coevolution: Has *Passiflora Adenopoda* Won the Selectional Race with Heliconine Butterflies?" *Science* 172 (1971): 585–586.

47. Lawrence B. Slobodkin, *The Growth and Regulation of Animal Populations* (New York: Holt, Reinhart and Winston, 1961); Vero C. Wynne-Edwards, *Animal Dispersion in Relation to Social Behavior* (Edinburgh: Oliver and Boyd, 1962); Vero C. Wynne-Edwards, "Intergroup Selection in the Evolution of Social Systems," *Nature* 100 (1963): 623–628; Vero C. Wynne-Edwards, "Population Control in Animals," *Scientific American* 211 (1964): 68–74.

48. Eric Klopfer, "Evolution of Intermediate Exploitation Rates in Exploiter-Victim Systems" (Ph.D. diss., University of Wisconsin at Madison, 1997).

49. Kinzig and Harte, "Selection of Microorganisms in a Spatially Heterogeneous Environment."

50. Nowak and May, "Evolutionary Games and Spatial Chaos"; Durrett and Levin, "The Importance of Being Discrete and (Spatial)."

51. Robert H. Whittaker, *Communities and Ecosystems,* 2nd ed. (New York: Macmillan, 1970).

Chapter Seven

1. National Research Council, Ocean Studies Board, *Sustaining Marine Fisheries* (Washington, D.C.: National Academy Press, 1999).

2. Cedar Creek is one of the long-term experimental research sites sponsored by the National Science Foundation as part of an effort to understand the structure and functioning of diverse ecosystems.

3. David Tilman and John A. Downing, "Biodiversity and Stability in Grasslands," *Nature* 367 (1994): 363–365.

4. See, for example, David A. Wardle, Olle Zackrisson, Greger Hörnberg, and Christiane Gallet, "The Influence of Island Areas on Ecosystem Properties," *Science* 277 (1997): 1296–1299; David U. Hooper and Peter M. Vitousek, "The Effects of Plant Composition and Diversity on Ecosystem Processes," *Science* 277 (1997): 1302–1305; Jill McGrady-Steed, Patricia M. Harris, and Peter J. Morin, "Biodiversity Regulates Ecosystem Predictability," *Nature* 390 (1997): 162–165.

5. Shahid Naeem, Lindsey J. Thompson, Sharon P. Lawler, John H. Lawton, and Richard M. Woodfin, "Declining Biodiversity Can Alter the Performance of Ecosystems," *Nature* 368 (1994): 734–737; Shahid Naeem, Lindsey J. Thompson, Sharon P. Lawler, John H. Lawton, and Richard M. Woodfin, "Biodiversity and Ecosystem Functioning: Empirical Evidence from Experimental Microcosms," *Philosophical Transactions of the Royal Society, London B 347* (1995): 249–262.

6. Tilman and Downing, "Biodiversity and Stability in Grasslands."

7. Daniel F. Doak, David Bigger, Elaine K. Harding, Michelle A. Marvier, Rachel E. O'Malley, and Diane Thomson, "The Statistical Inevitability of Stability-Diversity Relationships in Community Ecology," *American Naturalist* 151 (1998): 264–276.

8. J. Leps, J. Osbornova-Kosinova, and Marcel Rejmanek, "Community Stability, Complexity, and Species Life-History Strategies," *Vegetatio* 50 (1982): 53–63; C. W. MacGillivray, J. P. Grime, and the Integrated Screening Programme Team, "Testing Predictions of the Resistance and Resilience of Vegetation Subjected to Extreme Events," *Functional Ecology* 9 (1995): 640–649; J. P. Grime, "Biodiversity and Ecosystem Function: The Debate Deepens," *Science* 277 (1997): 1260–1261.

9. B. R. Trenbath, "Neighbour Effects in the Genus Avena II: Comparison of Weed Species," *Journal of Applied Ecology* 12 (1975): 189–200; Michael A. Huston, *Oecologia* 110 (1997): 449–460.

10. Grime, "Biodiversity and Ecosystem Function"; David Tilman, Johannes Knops, David Wedin, Peter Reich, Mark Ritchie, and Evan Siemann, "The Influence of Functional Diversity and Composition on Ecosystem Processes," *Science* 277 (1997): 1300–1302.

11. Grime, "Biodiversity and Ecosystem Function."

12. Robert V. O'Neill, Donald L. DeAngelis, Jack B. Waide, and Timothy F. H. Allen, *A Hierarchical Concept of Ecosystems,* Monographs in Population Biology, vol. 23 (Princeton, N.J.: Princeton University Press, 1986).

13. Arthur Koestler, *The Ghost in the Machine* (New York: Macmillan, 1967); Arthur Koestler, "Beyond Atomism and Holism: The Concept of the Holon," in Arthur Koestler and John R. Smythies, eds., *Beyond Reductionism* (London: Hutchinson 1969), 192–232.

14. Robert T. Paine, "Food Webs: Linkages, Interaction Strength, and Community Infrastructure," *Journal of Animal Ecology* 49 (1980): 667–685.

15. Koestler, *The Ghost in the Machine;* E. G. Leigh and T.E. Rowell, "The evolution of mutualism and other forms of harmony at various levels of biological organization," *Ecologie* 26 (1995): 131–158.

16. Herbert A. Simon, "The Architecture of Complexity," *Proceedings of the American Philosophical Society* 106 (1962): 467–482; see also Herbert A. Simon, *The Science of Artificial Life* (Cambridge, Mass.: MIT Press, 1969).

17. Herbert A. Simon, personal communication, August 1998.

18. O'Neill et al., *A Hierarchical Concept of Ecosystems.*

19. Stephen A. Forbes, *Ecological Investigations of Stephen Alfred Forbes* (1977; reprint, New York: Arno Press, 1977).

20. The subject is indebted to Stuart Pimm for discovering this work, to Joel Cohen for making it more broadly available, and to Claudia Jacobi for

translating it; Lorenzo Camerano, "Dell'equilibrio dei viventi mercè la reciproca distruzione" (On the equilibrium of living beings by means of reciprocal destruction), *Atti della Reale Accademia delle Scienze di Torino* 15 (1880): 393–414; Joel E. Cohen, "Lorenzo Camerano's Contribution to Early Food Web Theory," in Simon A. Levin, ed., *Frontiers in Mathematical Biology*, vol. 100 (Heidelberg: Springer, 1994), 351–359.

21. Camerano, "Dell'equilibrio dei viventi mercè la reciproca distruzione."

22. Charles Elton, *Animal Ecology* (New York: Macmillan, 1927).

23. Raymond L. Lindeman, "The Tropic-Dynamic Aspect of Ecology," *Ecology* 23 (1942): 399–413.

24. One of Elton's other major contributions, long before others recognized its importance, was the global threat to biodiversity caused by the transportation and introduction of such species from places where they belonged to places where they didn't; Charles S. Elton, *The Ecology of Invasions by Animals and Plants* (London: Methuen & Co., 1958).

25. Robert M. May, *Stability and Complexity in Model Ecosystems* (Princeton, N.J.: Princeton University Press, 1973).

26. Joseph H. Connell, "Effect of Competition, Predation by *Thais Lapillus*, and Other Factors on Natural Populations of the Barnacle *Balanus Balanoides*," *Ecological Monographs* 31 (1961): 61–104; Joseph H. Connell, "The Influence of Interspecific Competition and Other Factors on the Distribution of the Barnacle *Chthamalus Stellatus*," *Ecology* 42 (1961): 710–723; Robert T. Paine, "Food Web Complexity and Species Diversity," *American Naturalist* 100 (1966): 65–75.

27. Oscar H. Paris, "Some Quantitative Aspects by Muricid Snails on Mussels in Washington Sound," *Veliger* 2 (1960): 41–47.

28. Paine, "Food Webs."

29. Robert T. Paine, *Marine Rocky Shores and Community Ecology: An Experimentalist's Perspective* (Oldendorf/Luhe, Germany: Ecology Institute, 1994).

30. Jane Lubchenco, "Plant Species Diversity in a Marine Intertidal Community: Importance of Herbivore Food Preference and Algal Competitive Abilities," *American Naturalist* 112 (1978): 23–39; Bruce A. Menge, "Organization of the New England Rocky Intertidal Community: Role of Predation, Competition, and Environmental Heterogeneity," *Ecological Monographs* 46 (1976): 355–393; Bruce A. Menge, "Indirect Effects in Marine Intertidal Interaction Webs: Patterns and Importance," *Ecological Monographs* 65 (1995): 21–74; Paul K. Dayton, "The Structure and Regulation of Some South American Kelp Communities," *Ecological Monographs* 55 (1985): 447–468. See also Simon A. Levin and Robert T. Paine, "Distur-

bance, Patch Formation, and Community Structure," *Proceedings of the National Academy of Science* 71 (1974): 2744–2747; Robert T. Paine and Simon A. Levin, "Intertidal Landscapes: Disturbance and the Dynamics of Pattern," *Ecological Monographs* 51 (1981): 145–178.

31. J. A. Estes and J. F. Palmisano, "Sea Otters: Their Role in Structuring Near-shore Communities," *Science* 185 (1974): 1058–1060.

32. Paine, *Marine Rocky Shores and Community Ecology.*

33. Mary E. Power et al., "Challenges in the Quest for Keystones," *BioScience* 46 (1996): 609–620.

34. Anthony R. E. Sinclair, "The Eruption of the Ruminants," in Anthony R. E. Sinclair and M. Norton-Griffiths, eds., *Serengeti: Dynamics of an Ecosystem* (Chicago: University of Chicago Press, 1979), 82–103; Samuel J. McNaughton, "The Propagation of Disturbance in Savannas Through Food Webs," *Journal of Vegetation Science* 3 (1992): 301–314.

35. Paine, *Marine Rocky Shores and Community Ecology.*

36. Kari J. Laine and Pekka Niemela, "The Influence of Ants on the Survival of Mountain Birches During an Oporinia Autumnata *(Lepidoptrera, Geometridae)* Outbreak," *Oecologia* 47 (1980): 37–42.

37. Richard B. Root, "The Niche Exploitation Pattern of the Blue-gray Gnatcatcher," *Ecological Monographs* 37 (1967): 317–350.

38. There is a strong confluence of views in the approaches of Paine and Root, perhaps tracing back in part to their days together at the University of Michigan. Root was a precocious undergraduate at Michigan while Paine was completing his doctoral work with Fred Smith; Michigan was then in its heyday, perhaps the leading place to study ecology at that time. Root and Paine have been among my closest friends and colleagues for thirty years, and my own perspectives have been heavily influenced by their insights.

39. Lumping and splitting represent a fundamental dichotomy of views even in this endeavor.

40. Indeed, there is no reason to restrict ourselves to species. For some purposes, it may be most sensible to lump together, for example, the juvenile stages of a number of fish species as one functional group, and the adult stages as another, because of similarities in feeding habits.

41. Robert S. Steneck and Megan N. Dethier, "A Functional Group Approach to the Structure of Algal-Dominated Communities," *Oikos* 69 (1994): 476–498.

42. M. E. Hay, "Species as 'noise' in community ecology: do seaweeds block our view of the kelp forest?," *Trends in Ecology & Evolution* 9 (1994): 414–416.

43. David Raffaelli and Stephen J. Hall, "Compartments and Predation in an Estuarine Food Web," *Journal of Animal Ecology* 61 (1992): 551–560;

D. G. Raffaelli and S. J. Hall, "Assessing the Relative Importance of Trophic Links in Food Webs," in Gary A. Polis and Kirk O. Winemiller, eds., *Food Webs: Integration of Patterns and Dynamics* (New York: Chapman & Hall, 1996), 185–191.

44. Howard T. Odum, "Trophic Structure and Productivity of Silver Springs, Florida," *Ecological Monographs* 27 (1957): 55–112.

45. Robert T. Paine, "On Food Webs: Road Maps of Interactions or Grist for Theoretical Development?" *Ecology* 69 (1988): 1648–1654; Gary A. Polis, "Complex Trophic Interaction in Deserts: An Empirical Critique of Food-Web Theory," *American Naturalist* 138 (1991): 123–155; Gary A. Polis, "Food Webs, Trophic Cascades, and Community Structure," *Australian Journal of Ecology* 19 (1994): 121–136.

46. Stuart L. Pimm, *Food Webs* (New York: Chapman & Hall, 1982); Stuart Pimm, John Lawton, and Joel Cohen, "Food Web Patterns and Their Consequences," *Nature* 350 (1991): 669–674.

47. Pimm, *Food Webs.*

48. Robert H. MacArthur, "Patterns of Communities in the Tropics," *Biological Journal of the Linnean Society* 1 (1969): 19–30; Dolph Schluter and Robert E. Ricklefs, "Species Diversity: An Introduction to the Problem," in Robert E. Ricklefs and Dolph Schluter, eds., *Species Diversity in Ecological Communities: Historical and Geographical Perspectives* (Chicago: University of Chicago Press, 1993), 1–10.

49. Recall, too, that studies such as Tilman's show that there are reciprocal influences of diversity on productivity. We are just beginning to understand these effects, which represent an important part of understanding the correlations described in this paragraph.

50. Michael L. Rosenzweig and Zvika Abramsky, "How Are Diversity and Productivity Related?" in Ricklefs and Schluter, *Species Diversity in Ecological Communities.*

51. This would seem to have implications for economic markets as well, but to my knowledge, the connections have not yet been explored.

52. John J. Ewel, Maria J. Mazzarino, and Cory W. Berish, "Tropical Soil Fertility Changes Under Monocultures and Successional Communities of Different Structure," *Ecological Applications* 1 (1991): 289–302; Peter M. Vitousek and David U. Hooper, "Biological Diversity and Terrestrial Ecosystem Biogeochemistry," in E. D. Schulze and H. A. Mooney, eds. *Biodiversity and Ecosystem Function* (New York: Springer, 1994), 3–14.

53. Crawford (Buzz) S. Holling, "Resilience and Stability of Ecological Systems," *Annual Review of Ecology and Systematics* 4 (1973): 1–23.

54. Stephen R. Carpenter and James F. Kitchell, eds., *The Trophic Cascade in Lakes* (Cambridge: Cambridge University Press, 1993).

55. Paine, "Food Webs"; Robert M. May, "Will a Large Complex System Be Stable?" *Nature* 238 (1972): 413–414; Roy E. Plotnick and Michael L. McKinney, "Ecosystem Organization and Extinction Dynamics," *Palaios* 8 (1993): 202–212.

56. Duncan J. Watts and Steven H. Strogatz, "Collective Dynamics of 'Small-World' Networks," *Nature* 393 (1998): 440–442.

57. Bad jokes, perhaps unfortunately, typically have small effects and hence are not contained.

58. Simon A. Levin et al., "Resilience in Natural and Socioeconomic Systems," *Environment and Developmental Economics* 3 (1998): 225–236.

59. Carpenter and Kitchell, *The Trophic Cascade in Lakes.*

Chapter Eight

1. Percy Bysshe Shelley, *Prometheus Unbound* (1819).

2. Ralph E. Gomory, "The Known, the Unknown, and the Unknowable," *Scientific American* 276 (1995): 120.

3. For a discussion of leaf forms in variable environments, see Thomas Givnish, "On the Adaptive Significance of Leaf Form," in O. T. Solbrig et al., eds., *Topics in Plant Population Biology* (New York: Columbia University Press, 1979), 375–407.

4. Brian F. Chabot and Jean Fincher Chabot, "Effects of Light and Temperature on Leaf Anatomy and Photosynthesis in *Fragaria Virginiana*," *Oecologia* 26 (1977): 363–377; Louis J. Gross, "Photosynthetic Dynamics in Varying Light Environments: A Model and Its Applications to Whole Leaf Carbon Gain," *Ecology* 63 (1982): 84–93.

5. An exception is the game Extinction: The Game of Ecology, the creation of my colleague Steve Hubbell (2nd ed., Burlington, N.C.: Carolina Biological Supply, 1978).

6. Shay Gueron and Simon A. Levin, "The Dynamics of Group Formation," *Mathematical Biosciences* 128 (1995): 243–264.

7. Per Bak, *How Nature Works* (New York: Springer, 1996).

8. But see a strong dissent by the distinguished paleontologist David Raup, "A Breakthrough Book?" *Complexity* 2 (1997): 30–32.

9. Stephen Jay Gould and Niles Eldredge, "Punctuated Equilibria: The Tempo and Mode of Evolution Reconsidered," *Paleobiology* 3 (1977): 115–151.

10. Robert H. MacArthur and Edward O. Wilson, *The Theory of Island Biogeography*, Monographs in Population Biology, vol. 1 (Princeton, N.J.: Princeton University Press, 1967).

11. Robert M. May and Robert H. MacArthur, "Niche Overlap as a Function of Environmental Variability," *Proceedings of the National Academy of Science, USA* 69 (1972): 1109–1113.

12. Bak, *How Nature Works,* 155.

13. Stuart A. Kauffman, *At Home in the Universe* (New York: Oxford University Press, 1995). See also Stuart A. Kauffman, *The Origins of Order* (New York: Oxford University Press, 1993); C. J. Langton, ed., *Artificial Life*, Santa Fe Institute Studies in the Sciences of Complexity, vol. 6 (Redwood City, Calif.: Addison Wesley, 1989).

14. Kauffman *(At Home in the Universe)* argues that "[t]he reason complex systems exist on, or in the ordered regime near, the edge of chaos is because evolution takes them there" (90). Kauffman does not make a clear distinction between self-organized criticality and the edge of chaos, a quite different concept technically.

15. Ibid., 235.

16. Bak and Sneppen do introduce a competitive mechanism that represents a form of evolution, but it is drastically oversimplified. Per Bak and Kim Sneppen, "Punctuated Equilibrium and Criticality in a Simple Model of Evolution," *Physical Research Letters* 24 (1993): 4083–4086. See also Per Bak, Henrik Flyvbjerg, and Kim Sneppen, "Can We Model Darwin?" *New Scientist* 12 (1994): 36; Kim Sneppen, Per Bak, Henrik Flyvbjerg, and Mogens Høth Jensen, "Evolution as a Self-organized Critical Phenomenon," *Proceedings of the National Academy of Science, USA* 92 (1995): 5209–5213.

17. B. B. Mandelbrot, *Fractals: form, chance, and dimension* (San Francisco: W. H. Freeman & Co., 1977).

18. Simon A. Levin and Linda Buttel, *Measures of Patchiness in Ecological Systems* (Ithaca, N.Y.: Ecosystems Research Center, Cornell University, 1997); Simon A. Levin, "The Problem of Pattern and Scale in Ecology," *Ecology* 73 (1992): 1943–1967.

19. Bak and Sneppen, "Punctuated Equilibrium and Criticality in a Simple Model of Evolution"; Ricardo V. Solé and Susanna C. Manrubia, "Are Rainforests Self-organized in a Critical State?" *Journal of Theoretical Biology* 173 (1995): 31–40; K. Moloney and Simon A. Levin, "The Effects of Disturbance Architecture on Landscape-Level Population Dynamics," *Ecology* 772 (1996): 375–394.

20. Stephen W. Pacala and Simon A. Levin, "Biologically Generated Spatial Pattern and the Coexistence of Competing Species," in David Tilman and Peter Kareiva, eds., *Spatial Ecology: The Role of Space in Population Dynamics and Interspecific Interactions* (Princeton, N.J.: Princeton University Press, 1997), 204–232.

21. John T. Bonner, *Life Cycles* (Princeton, N.J.: Princeton University Press, 1993).

22. Luigi L. Cavalli-Sforza and Marcus J. Feldman, *Cultural Transmission and Evolution,* Monographs in Population Biology (Princeton, N.J.: Princeton University Press, 1981).

23. Ibid.

24. Herbert A. Simon, "Organizations and Markets," *Journal of Economic Perspectives* 5 (1991): 25–44.

Chapter Nine

1. Garrett Hardin, "The Tragedy of the Commons," *Science* 162 (1968): 1243–1248.

2. Gene M. Grossman and Alan B. Krueger, "Environmental Impacts of a North American Free Trade Agreement," in Peter Garber, ed., *The U.S.-Mexico Free Trade Agreement* (Cambridge, Mass.: MIT Press 1993), 165–177.

3. Kenneth Arrow et al., "Economic Growth, Carrying Capacity, and the Environment," *Science* 268 (1995): 520–521.

4. Rachel Carson, *Silent Spring* (Boston: Houghton Mifflin, 1962).

5. A case in point involves our efforts to manage disease through the use of antibiotics: the stronger our efforts have been, the stronger have been the selective forces for the evolution of new strains of organisms containing resistance to those antibiotics.

6. Edward O. Wilson, *The Diversity of Life* (London: Penguin Press, 1992).

7. Robert M. May, "How Many Species Inhabit the Earth?" *Scientific American* 267 (1992): 42–48; Robert M. May, "Biological Diversity: Differences Between Land and Sea," *Philosophical Transactions of the Royal Society of London B* 343 (1994): 105–111.

8. Partha Dasgupta, *An Inquiry into Well-being and Destitution* (Oxford: Clarendon Press, 1993).

9. Scott Barrett, "The Problem of Global Environmental Protection," *Oxford Review of Economic Policy* 6 (1990): 68–79; Simon A. Levin et al., "Resilience in Natural and Socioeconomic Systems," *Environment and Developmental Economics* 3 (1998): 225–236.

10. Crawford S. Holling, ed., *Adaptive Environmental Assessment and Management* (New York: Wiley, 1978). Much of Holling's work on adaptive management was developed when he was at the University of British Columbia, and later as director of the International Institute for Applied Sys-

tems Analysis (IIASA) in Laxenburg, Austria. In this, he had the collaboration of many distinguished colleagues, most notably Carl Walters, Ray Hilborn, Don Ludwig, the late Mike Fiering, and my old mentor, George Dantzig.

11. Ralph E. Gomory, "The Known, the Unknown, and the Unknowable," *Scientific American* 272 (1995): 120.

12. Syukuro Manabe and Ronald J. Stouffer, "Two Stable Equilibria of a Coupled Ocean-Atmosphere Model," *Journal of Climate* 1: 841–866.

13. Wallace S. Broeker, Dorothy M. Peteet, and David Rind, "Does the Ocean-Atmosphere System Have More Than One Stable Mode of Operation?" *Nature* 315 (1985): 21–26.

14. William D. Nordhaus, *Managing the Global Commons: The Economics of Climate Change* (Cambridge, Mass.: MIT Press, 1994).

15. Per Bak, *How Nature Works* (New York: Springer, 1996).

16. D. J. Watts and S. H. Strogatz, "Collective Dynamics of 'Small-World' Networks," *Nature* 393 (1998): 440–442.

17. Ian Player, *The White Rhino Saga* (New York: Stein and Day, 1973).

18. Graciela Chichilnisky and Geoffrey Heal, "Economic Returns from the Biosphere," *Nature* 391 (1998): 629–630.

19. Dasgupta, *An Inquiry into Well-being and Destitution.*

20. Matt Ridley and Bobbi S. Low, "Can Selfishness Save the Environment?" *Atlantic Monthly* (September 1993): 76–84.

21. Ibid.; Colin W. Clark, "The Economics of Overexploitation," *Science* 181 (1973): 630–634.

22. Chichilnisky and Geoffrey Heal, "Economic Returns from the Biosphere."

23. Geoffrey M. Heal, "Markets and Sustainability," in Richard B. Stewart, ed., *Environmental Governance* (Cambridge: Cambridge University Press, 1999).

24. Chichilnisky and Heal, "Economic Returns from the Biosphere."

25. Charlie Pye-Smith and Grazia Feyerabend, with Richard Sandbrook, *The Wealth of Communities* (London: Earthscan, 1994); Levin et al., "Resilience in Natural and Socioeconomic Systems." For informative and complementary perspectives on property rights and human societies, see Fikret Berkes, "Social Systems, Ecological Systems, and Property Rights," in Susan S. Hanna, Carl Folke, and Karl-Goran Mäler, eds., *Rights to Nature* (Washington, D.C.: Island Press, 1998), 87–107; and Fikret Berkes and Carl Folke, eds., *Linking Social and Ecological Systems: Management Practices and Social Mechanisms for Building Resilience* (Cambridge: Cambridge University Press, 1998).

26. Ridley and Low, "Can Selfishness Save the Environment?"

27. Elinor Ostrom, *Governing the Commons: The Evolution of Institutions for Collective Action* (Cambridge: Cambridge University Press, 1990); Elinor Ostrom, *Crafting Institutions for Self-governing Irrigation Systems* (San Francisco: Institute for Contemporary Studies Press, 1992).

28. Ridley and Low, "Can Selfishness Save the Environment?"

29. Abram Chayes and Antonia H. Chayes, *The New Sovereignty* (Cambridge, Mass.: Harvard University Press, 1995); Levin et al., "Resilience in Natural and Socioeconomic Systems."

30. Scott Barrett, "The Problem of Global Environmental Protection," *Oxford Review of Economic Policy* 6 (1990): 68–79; Levin et al., "Resilience in Natural and Socioeconomic Systems."

31. Robert M. Solow, "Sustainability: An Economist's Perspective," *Eighteenth J. Seward Johnson Lecture in Marine Policy* (Woods Hole, Mass.: Marine Policy Center, Woods Hole Oceanographic Institution, 1991).

32. *Debt-for-nature swaps* are a brilliant scheme of buying at discounts the unpaid debts of impoverished nations and forgiving them in exchange for the preservation of land in an undeveloped status. They were the brainchild of the conservation biologist Thomas Lovejoy.

33. See, for example, Mark W. Brunson, "Social Dimensions of Boundaries: Balancing Cooperation and Self-interest," in Richard L. Knight and Peter B. Landres, eds., *Stewardship Across Boundaries* (Washington, D.C.: Island Press, 1998), 65–86; and other related papers.

34. Joseph Herman Hertz, *The Pentateuch and Haftorahs,* 2nd ed. (London: Soncino Press, 1963), 563.

GLOSSARY

abiotic Not biotic

adaptation The evolution of traits through natural selection; also, a trait that has arisen through natural selection, based on differential fitness

adaptive landscape A pictorial representation of the relationship between fitness and genotype, or between fitness and the genetic composition of a population; developed by Sewall Wright

adaptive probing A management device, characterized by exploration of diverse strategies, to improve information about alternatives

aggregation Lumping diverse objects together into a single category

allele A form or realization of a gene

allelopathy A competitive device involving the production of chemicals to harm others

altruist An individual that performs an activity at the apparent sacrifice of its immediate benefit

amino acid Building block for proteins and other polypeptides; composed of an amino acid group and a carboxylic acid group

anthropic principle The notion that, if life can exist only under a limited range of conditions, the world in which we exist must satisfy those conditions; hence, there is an observer bias

apostatic selection Natural selection on prey characteristics as a result of predators switching from rare to common types of prey

artificial selection Genetic change within populations caused by the actions of breeders

aspect diversity The diversity within a population for any morphological characteristics, such as color or wing patterns

autotrophic Photosynthetic, or otherwise able to manufacture food from inorganic materials

bacteriocin Chemical produced by bacteria to poison other bacteria

bell curve The symmetric distributional curve characteristic of the normal distribution

biodiversity The diversity of biological types at all levels of organization, including heterogeneity within and among populations

biological diversity See *biodiversity*

biomass The total weight of an individual, a population, or a collection of populations

biosphere The Earth as a system, including its biological, physical, and chemical dimensions

biota The totality of organisms within a particular region; the flora and fauna together

birth rate Number of births per unit of time

cellular automata Simplified mathematical models of spatial interactions, in which sites or cells on a landscape are assigned a particular state, which changes according to specific rules conditioned on the states of neighboring cells

cellular slime mold Various fungi whose vegetative structures are aggregations of amoeboid cells

chaos Complex dynamical behavior characterized by lack of dominant periodicity, and by great sensitivity to initial conditions

character displacement The evolutionary divergence of similar traits among coexisting competitors

chromosome The site of genes; linear structure within cell nucleus providing the location of the DNA

climax The end result of succession within ecological communities

coevolution The evolution of two interacting species, under mutual influence on one another

coexistence The existence of two or more species together within the same habitat

colicinogenic Producing colicin, a bacterial poison

commensalism The condition of two organisms living together without harming one another

community The collection of organisms and populations of various species within a particular habitat

community respiration Transformation of oxygen into carbon dioxide and other products of oxidation by all the plants in a community

competitive exclusion The elimination of one or more of a set of competing species, through the effects of competition

competitive release The numerical increase in abundance of a species due to the elimination of a competitor

competitor An individual or species that competes for resources with another

complex adaptive system A system composed of a heterogeneous assemblage of types, in which structure and functioning emerge from the balance between the constant production of diversity, due to various forces, and the winnowing of that diversity through a selection process mediated by local interactions

complexity Technically, an interconnected network of components that cannot be described by a few rules; generally manifest in structure, order and functioning emerging from the interactions among diverse parts

conservation biology The science and practice of preservation of species and habitats

constancy Not changing over time

contact process A particular form of an interacting particle system in which sites may be occupied or not by one of a variety of types

convergent evolution The evolution of similar features among noninteracting populations in different habitats, in response to similar environmental conditions and challenges

correlation Statistical measure of relatedness of events

correlation length The distance or time beyond which events are not related

criticality The condition describing the transition between qualitatively different states, such as solid/liquid or liquid/gas

crustose algae Thin and crusty algae

crypsis Camouflage patterns that allow organisms to conceal themselves against backgrounds

cybernetic Relating to a control process

Daisyworld A greatly simplified model (developed by Lovelock and collaborators) for how some homeostatic properties of the biosphere might have evolved

death rate Number of deaths per unit of time

debt-for-nature swap A device, the creation of Thomas Lovejoy, in which debts incurred by developing nations are bought at a discount, and then forgiven in exchange for agreements from those nations to preserve natural areas

denitrification The conversion of nitrate into nitrogen gases

dependent variable A quantity that depends upon other variables, that is, is a function of independent variables

development The progression of an organism from birth to adult; also, the economic growth and exploitation of a region

diffuse coevolution Coevolution among a number of species, no two of which interact strongly

diploid Genetic system in which each individual inherits chromosomes in pairs, one from each parent

direct methods Optimization methods that do not depend on computation of slopes

diversity Variety

DNA Deoxyribonucleic acid, whose various forms provide the alphabet underlying the genetic code; the fundamental building blocks of genes

ecosystem The integrated system involving the biological, physical, and chemical components of a habitat or region

ecotone The transition zone between two distinct kinds of ecological communities

Ecotron An experimental facility at Silwood Park, England, for the controlled study of ecological interactions

edge of chaos Technically, the state describing the transition of a system to chaos; used more loosely by some as equivalent to self-organized criticality, or even more loosely as the boundary between stultifying sameness and energizing turbulence

egoist An individual whose behavior benefits itself

embryogenesis The development of an organism from birth to embryo

emergence The development of properties in a self-organizing system, as the outgrowth of interactions between components

endemic A species that is unique to a particular area

endogenous Arising from within

equilibrium A sustained constant state in a mathematical model

etiology The cause of a disease

euphotic zone Upper layer in water bodies, defined by where photosynthesis exceeds respiration

evolution Change in the distribution of characteristics within a population, and of the distribution of populations, through the actions of natural selection, environmental change, mutation and recombination, chance factors, and other mechanisms

evolutionarily stable strategy A strategy that, once established in a community, cannot be displaced

exogenous Arising from without

extinction The permanent disappearance of a species from a habitat, or from the biosphere

extrinsic See *exogenous*

Farenholz's rule The rule that parasite speciation patterns often conform to the speciation patterns of their hosts

fauna The totality of animals or types of animals within a particular region

feedback Influence on a system component mediated by changes induced by that component

feedback loop A chain of influence of any length, in which a system component is influenced indirectly by changes it has induced

fitness The measure of the survival and reproductive success of an organism, or of a type of organism

flora The totality of plants or types of plants within a particular region

flow Movement from one place to another

flushing rate The rate at which liquids that enter a region are removed from it

food web The network of feeding relationships among the species in a region

fractal A self-similar structure whose geometrical and topological features are recapitulated in miniature on finer and finer scales

frequency dependence The influence of the distribution of types within a population on the fitnesses and evolution of all types

function A mathematical relationship between variables; also, ecologically synonomous with purpose or role

functional group A group of species performing similar functions within an ecosystem

fundamental biota The totality of species that could exist within a region

fundamental niche The range of conditions under which a species could exist

fundamental theorem of natural selection Sir Ronald Fisher's mathematical statement that the rate of change of a character under the influence of natural selection is proportional to the selection coefficient, and to the additive portion of genetic variation

Gaia The hypothesis of James Lovelock and Lynn Margulis that the biosphere is a homeostatic, cybernetic system in which the biota plays a fundamental role in regulating physical and chemical processes

game theory The theory describing the interactions between diverse strategies for playing various games; especially the search for optimal strategies

gene The fundamental unit of natural selection

genetic code The rules that translate the information encoded in genes into the production of amino acids and that therefore determine heredity

genetic variation The genetically based variation within a population

genotype A genetic type; that is, a type of organism as determined by its genetic characteristics

guild A group of species exhibiting similar strategies for resource acquisition within a community

habitat The environment in which an organism, population, or community exists

haplo-diploid The genetic system in which one sex arises from unfertilized eggs so that, for example, males are haploid and females are diploid

haploid The genetic system in which individuals receive only a single copy of each gene, hence from a single parent

heterogeneous Comprising dissimilar types

hierarchical organization Arranged or classified according to rank

hierarchy Hierarchical organization

holon A subset of a system involving components that interact relatively strongly among themselves, and relatively weakly with others

holothurian Sea cucumber

homeostasis The maintenance of constant state

homogeneous Comprising similar types

horizontal transfer The transfer of genetic material from one organism to another, exclusive of the hereditary process

host An organism that houses and provides nutrients to another, especially through a parasitic process

Hymenoptera Order of insects including the bees, ants, and wasps

independent variable A quantity that varies freely and upon which other variables depend

individualistic The notion that species are distributed on the landscape individually in terms of their relation to the environment and cannot be grouped easily into distinct characteristic communities of species with common distributions

industrial melanism The appearance of dark coloration among moths in sooty industrial areas as the result of natural selection for crypsis

interacting particle systems Stochastic cellular automata; in particular, spatial stochastic process models

intrinsic See *endogenous*

invertebrate Lacking a backbone

invisible hand The notion of the economist Adam Smith that efficient utilization of resources arises from a free market of individuals pursuing their own selfish agendas

island biogeography The theory developed by Robert MacArthur and Edward Wilson to describe the mechanisms and patterns characterizing the assembly of island biotas

Iterated Prisoner's Dilemma (IPD) A game in which *Prisoner's Dilemma* is played repeatedly, with memory

iteroparity Reproducing multiple times in a lifetime

JABOWA The original forest growth simulator, developed by Daniel Botkin and his colleagues
keystone predator A top predator species whose removal from a community would engender qualitative changes in community organization
keystone species Any species whose removal from a community would engender qualitative changes in community organization
Law of Large Numbers The mathematical principle, developed by Jacob Bernoulli, that the most likely fraction of independent trials that will exhibit a given behavior is the same as the probability of that behavior on a given trial, and that the probability of observing a given amount of deviation from that expectation decreases toward zero as the number of trials increases without bound
lichen A mutualistic association of a fungus and an alga
life zone Vegetation zone as specified by environmental characteristics such as climate and moisture
limnology The scientific study of lakes, ponds, and streams
linear programming A set of mathematical methods for finding the optima of linear functions under linear constraints
lognormal A statistical distribution in which the logarithm of the quantity of interest has a normal distribution
macroalgae Large algae, such as kelps
macroscopic On the broadest scale of a distribution or a process
mesoscopic On the medium scale of a distribution or a process
meta-agent Aggregate of agents
meta-community An ensemble of communities
meta-population An ensemble of populations, generally of a single species
microenvironment A subset of the environment, for example, one characterized by particular kinds of resources or soil properties
microscopic On the smallest scale of a distribution or a process
model A mathematical or physical simplification of a system
mutation The spontaneous change of a gene from one form to another
mutualism The reciprocally beneficial interactions between mutualists
mutualist An individual or species whose actions benefit an individual or individuals of another species
mycorrhizae Soil fungi that form mutualistic associations with plants, attaching to roots and transforming minerals into forms usable by the plants
myxoma A virus that is pathogenic to certain rabbit populations; used as a pest control device in Australia and Europe
natural selection The process that translates differential success in survival and reproduction into evolutionary change; Herbert Spencer termed it the "survival of the fittest"

negative feedback Feedback that tends to counteract a process
niche The role of a species within its community
NIMBY Acronym for "not in my back yard"
nitrification The oxidation of ammonia to produce nitrates that plants can easily assimilate
nitrogen cycle The cycling of nitrogen through the processes of nitrogen fixation, nitrification, and denitrification
nitrogen fixation The conversion of gaseous nitrogen into other compounds, such as ammonium and ammonia, making it available for oxidation for energy or assimilation by some organisms
nonlinear Disproportionate in cause and effect
normal distribution A statistical distribution of a specific symmetric nature, representing the outcomes of a large number of basic random processes
ontogeny The sequence of stages in the development of an organism from birth to adult
opportunistic A species that exploits transient environmental opportunities
optimality Condition of being best
optimization The search for the best solution among alternatives, or the extreme value of a variable or function
parasite An individual or species that exploits another individual (host) or individuals, especially of a different species
path dependency The dependence of the outcome of a dynamical process on historical events
pathogen An organism that causes disease
pattern formation The emergence of large-scale patterns from local processes and interactions
percolation Mathematical theory of spread on cellular landscapes
periodic Characterized by repetition of patterns in fixed intervals of space or time
phase transition Transformation of a substance or a system from one form to a qualitatively different one
photic zone The upper zone of a body of water, as defined by the depth to which sunlight can penetrate
photosynthesis Conversion of light energy into chemical energy and synthesis of organic compounds into inorganic compounds, by green plants
phytoplankton Tiny drifting plants, especially algae, of the upper layer of water bodies
plankton Tiny drifting organisms of the upper layer of water bodies
plasmid Extrachromosomal circular DNA
polypeptide chain Chain of amino acids

population The totality of organisms of a species within a particular region

population genetics The study of genetic changes within a population

positive feedback Feedback that tends to reinforce a process

power law A relationship between two variables such that one is proportional to a power of the other

predator An animal that hunts and kills others for food

prey An animal that is hunted or killed for food by another

primary production The assimilation of sunlight energy by plants

primary productivity The rate of assimilation of sunlight energy by plants

primary succession The sequence of species colonizing an area after a novel disturbance, such as a lava flow

Prisoner's Dilemma A prototypical game theory problem involving two prisoners held in isolation; illustrative of the difficulties in achieving cooperative behavior

privatization The transfer of property rights to individuals and corporations

protein Large chain of amino acids

realized biota The assemblage of species found within a particular region

realized niche The range of conditions under which a species is found

recessive Forms of genes (alleles) that have no effect unless inherited from both parents

reciprocal altruism Mutually beneficial behavior in which an individual helps another because of expected reciprocal behavior

recombination The genetic reshuffling of material within an organism due to the crossing (over) of chromosomes and reassortment of genes

residence time The average time material spends in a system

resiliency Capability to resist qualitative shifts in behavior or characteristics

resistance The ability of a host to sustain itself against a parasite, or of a bacterium to sustain itself against an antibiotic

scale Extent or grain

schema Building block for discovering rules; an internal model of the real world

schemata Plural of *schema*

sclerophylls Plants with hard leaves

sea star Starfish

search image The mental image a predator uses in searching for a prey

secondary succession The sequence of species recolonizing space after a common disturbance, such as a treefall or a fire

selection See *natural selection*

self-organization The development of structure and functioning on the basis of local interactions

self-organized criticality The tendency of large complex systems to self-organize to a self-maintaining critical state

shifting balance theory A fundamental theory, developed by Sewall Wright, of evolutionary change over heterogeneous environments

simplex method Method developed by George Bernard Dantzig for solving linear programming problems

simulated annealing Optimization methods that simulate the physical process of annealing by allowing the occasional acceptance of less attractive solutions or values

simulator A computer model that simulates some process

SORTIE A forest growth simulator developed by Stephen Pacala and his colleagues

speciation The process by which new species are formed

species-area curve The relationship between the area of a region and the number of species found there

stability The tendency to return to normative behavior, such as an equilibrium or limit cycle in a mathematical model

stochastic Governed by chance and unpredictability

stochastic cellular automata Cellular automata in which rates are stochastic

stochastic optimization A class of mathematical optimization tools that rely in part on the random generation of new variants of existing forms

succession The sequence of species colonizing a region, especially following a disturbance

superorganism The concept of the ecological community, the ecosystem, or the biosphere as an ecological and evolutionary unit

susceptibility A measure of the propensity of a host to succumb to a parasite, or of a bacterium to succumb to an antibiotic

sustainable stewardship Economic development of a region in ways that can be sustained

tight coevolution Coevolution among a number of species that interact strongly

tit for tat A strategy for cooperative behavior in which an individual, at each stage, plays the strategy last played by another against it

Tragedy of the Commons The notion, explicated by Garrett Hardin, that though we live in a Commons, individuals will practice selfish behaviors that are not in the common good and lead to overexploitation of resources

trait A characteristic of an organism

trait group Notion introduced by David Sloan Wilson of a group of individuals that share a common trait

traveling salesman problem Classic problem in operations research of finding the shortest path a traveling salesman could follow in order to visit each of a set of geographically distributed cities

triage An allocation system for medical or conservation practice that provides full treatment for individuals, species, or habitats of highest priority and none for those of lowest priority

trophic level Number of steps energy must take from the sun to reach a particular species, through a chain of consumption of one species by another

trophic pyramid The conceptualization of the trophic web in which higher trophic levels capture less energy

trophic web The network of energy flow in a community; see also *food web*

turf algae Algae that exhibit dense growth form

turnover rate The rate of species replacement in an island or other habitat

uncorrelated Independent

uptake Assimilation of nutrients from the environment

vector An organism or species that carries the agents of disease from one host organism to another

vertebrate Possessing a backbone

virulence Measure of the degree of pathogenicity of a parasite, or of the population level (titer) it can reach in a host organism

voter model A particular form of an interacting particle system in which every site is occupied by a specific type, whose identity changes according to rules conditioned on the states of neighboring cells

zooplankton Tiny drifting animals, especially fish larvae and invertebrates, of the upper layer of water bodies

INDEX